Dark Tourism Studies

This book provides original, innovative, and international tourism research that is embedded in interdisciplinary and transdisciplinary theoretical and methodological thought in the study of dark tourism.

It is almost 25 years since the idea of dark tourism was introduced and presented into the field of tourism studies. The impact of this idea was greater, which attracted a great deal of attention from different researchers and practitioners with a good range of disciplines and farther tourism studies. This edited volume aims to capture a glimpse of the types of cutting-edge thinking and academic research in the domain of dark tourism studies as well as encourage and advance theoretical, conceptual, and empirical research on dark tourism. This book also addresses several future research directions focusing on the experience and emotions of visitors at 'dark tourism' sites.

This book will be valuable reading for students, researchers, and academics interested in dark tourism. Other interested stakeholders including those in the tourism industry, government bodies, and community groups will also find this volume relevant.

The chapters in this book were originally published as a special issue of the *Journal of Heritage Tourism*.

Rami K. Isaac is Senior Lecturer in Tourism at the Breda University of Applied Sciences, the Netherlands. He was born in Palestine and did his undergraduate studies in the Netherlands and graduate studies in the UK, and has earned his PhD from the University of Groningen, the Netherlands. His research interests are in tourism development and management, critical theory, and political aspects of tourism. He published numerous articles and book chapters on tourism and political (in)stability, occupation, tourism and war, dark tourism, violence, and transformational tourism.

Dark Tourism Studies

Edited by
Rami K. Isaac

R Routledge
Taylor & Francis Group

LONDON AND NEW YORK

First published 2022
by Routledge
4 Park Square, Milton Park, Abingdon, Oxon, OX14 4RN

and by Routledge
605 Third Avenue, New York, NY 10158

Routledge is an imprint of the Taylor & Francis Group, an informa business

© 2022 Taylor & Francis

British Library Cataloguing-in-Publication Data
A catalogue record for this book is available from the British Library

ISBN13: 978-1-032-21099-5 (hbk)
ISBN13: 978-1-032-21100-8 (pbk)
ISBN13: 978-1-003-26672-3 (ebk)

DOI: 10.4324/9781003266723

Typeset in Minion Pro
by codeMantra

Publisher's Note
The publisher accepts responsibility for any inconsistencies that may have arisen during the conversion of this book from journal articles to book chapters, namely the inclusion of journal terminology.

Disclaimer
Every effort has been made to contact copyright holders for their permission to reprint material in this book. The publishers would be grateful to hear from any copyright holder who is not here acknowledged and will undertake to rectify any errors or omissions in future editions of this book.

Contents

Citation Information

The following chapters were originally published in the *Journal of Heritage Tourism*, volume 16, issue 4 (2021). When citing this material, please use the original page numbering for each article, as follows:

Chapter 7

Dark tourism as educational tourism: the case of 'hope tourism' in Fukushima, Japan
Kyungjae Jang, Kengo Sakamoto and Carolin Funck
Journal of Heritage Tourism, volume 16, issue 4 (2021) pp. 481–492

For any permission-related enquiries please visit:
http://www.tandfonline.com/page/help/permissions

Notes on Contributors

Bailey Ashton Adie is Research Fellow in the Faculty of Business, Law and Digital Technologies at Solent University, Southampton, UK. Her research interests include World Heritage tourism and management, sustainable heritage tourism for community development, second home tourism, tourism and disasters, tourism branding and marketing, film tourism, overtourism, and visitor management.

Paul Barron is Professor of Hospitality and Tourism Management at Edinburgh Napier University, UK. He is currently convenor of University Research Integrity Committee, has authored articles in the fields of hospitality and tourism, and served as Executive Editor of the *Journal of Hospitality and Tourism Management* for six years.

Carolin Funck obtained her PhD from Albert-Ludwigs University, Freiburg, Germany. She is Professor for Human Geography at Hiroshima University (Japan), Graduate School of Humanities and Social Sciences. Her research focuses on the development of tourism in Japan and sustainable island tourism.

Ilze Grinfelde is Mg.soc.sc and works as Lecturer at the Faculty of Social Sciences and as research assistant in the Institute of Social, Economic and Humanities research at the Vidzeme University of Applied Sciences, Valmiera, Latvia. Her research interests are dark tourism, niche tourism, and sustainable regional development.

Ker Ni (Vivienne) Heng Rigney is an academic at Edith Cowan University and Murdoch University, Perth, Australia. Her research collaborations include marketing and management, and more recently commemorative tourism.

Rami K. Isaac is Senior Lecturer in Tourism at the Breda University of Applied Sciences, the Netherlands. His research interests are in tourism development and management, critical theory, and political aspects of tourism. He published numerous articles and book chapters on tourism and political (in)stability, occupation, tourism and war, dark tourism, violence, and transformational tourism.

Kyungjae Jang is Associate Professor/Lecturer in the Graduate School of Humanities and Social Sciences at Hiroshima University, Japan. He holds a PhD and MA in tourism studies from Hokkaido University, Sapporo, Japan, and a BA from Korea University, Seoul, South Korea. Jang has conducted participatory research on popular culture-related tourism, and war memories and tourism.

James Kennell is Deputy Head of the Department of Marketing, Events and Tourism at the University of Greenwich, UK. He is also the Executive Director of the London Office of the International Tourism Studies Association and a Fellow of the Tourism Society. James carries out research into the public policy and political economy aspects of tourism, as well as into cultural and other non-corporate events in terms of their wider social and economic impacts.

Anna Leask is Professor of Tourism Management at Edinburgh Napier University, UK. She is on the Editorial Board for four international tourism journals and has been involved in the Scientific Committees for international conferences in Europe and USA.

Martin MacCarthy is an academic in the School of Business and Law at Edith Cowan University, Perth, Australia. His research interests include commemorative tourism, qualitative research methods, and experiential consumption.

Raymond Powell is Principal Lecturer in Tourism in the Department of Marketing, Events and Tourism at the University of Greenwich, UK. His academic interests currently centre on dark and heritage tourism, and especially the links between the two. Ray is currently fully engaged in teaching and research at all levels, with particular research interests in cultural tourism and heritage; dark tourism; museums; aspects of entrepreneurship; and employability.

Kengo Sakamoto works for JTB Corporation, Japan. He holds a BA in School of Integrated Arts and Sciences from Hiroshima University, Japan.

Esther J. Snell is Senior Lecturer in Criminology at Solent University, UK. Her research explores the construction, dissemination, and influence of printed representations of crime and justice. Her publications include *Criminal Justice During the Long Eighteenth Century: Theatre, Representation and Emotion* (Routledge, 2018).

Marianna Sigala is Professor at the University of South Australia, Adelaide, Australia, and Director of the Centre for Tourism & Leisure Management. She is an international authority in the field of technologies in tourism and services management with numerous awarded publications, research projects, and keynote presentations in international conferences. She is the Editor-In-Chief of the *Journal of Hospitality and Tourism Management*.

Effie Sterioupoulos is Lecturer of Event Management at William Angliss Institute, Melbourne (Australia). Her research activities focus on transformative experiences and their influence on consumer behaviour. She is interested in transformational experiences, especially when storytelling and emotions are involved. Effie also conducts research in experiential marketing where brands are used to influence consumer-based brand loyalty.

Linda Veliverronena holds a PhD in media and communication, and works as Assistant Professor at the Faculty of Social Sciences and as researcher in the Institute of Social, Economic and Humanities research at the Vidzeme University of Applied Sciences, Valmiera, Latvia. Her research interests focus on community engagement issues and cultural heritage studies.

Brianna Wyatt is Lecturer for Oxford Brookes University, UK. Her PhD thesis (Edinburgh Napier University, 2019) explored the influences on the design and management of interpretation at lighter dark visitor attractions in the UK and Ireland, with an aim to extend knowledge and contribute to professional practice.

Introduction to dark tourism

Rami K. Isaac

It is almost 25 years since the idea of dark tourism was announced, and presented into the field of tourism studies. The impact of this idea was greater, which attracted much of attention by different scholars and practitioners with a good range of disciplines and beyond tourism studies (Light, 2017). The idea of dark tourism has caught the eye of the media (Lennon, 2010) and a frequent subject for newspaper/magazine articles and television programmes (Light, 2017, p. 276). Once introduced, the idea appeared to offer a replacement understanding into the tourist experience, which thus took centre stage during a field that had traditionally been seen mainly as an economic activity (Ashworth & Isaac, 2015).

The first idea was initiated by Foley and Lennon (1996) using the term 'dark tourism'. The key focus was upon the identification of a new type of tourism destinations and tourism products that were quite different from the assumed pleasure conveyed by traditional tourism consumption. The effort was made to classify 'dark sites' with the key question being: Can we identify which tourism destinations are dark? Although this idea of identification was somewhat 'vague, and not clearly defined' (Hartmann, 2014, p. 167).

A different stream of thought was established somehow independently and advanced from the interest of Seaton (1996, p. 2) in battlefield tourism, and the fascination of tourists, at sites associated with death, with the introduction of the word 'thanatourism' which defined as 'actual, or symbolic encounters with death'. He addressed the question: Is it possible to recognize and classify dark tourism experiences? This was undoubtedly a shift in the focus from the site to the motivation, conceptually as well as empirically (Ashworth, 2004, Isaac et al., 2019; Isaac & Çakmak, 2014, 2016; Podoshen, 2013; Seaton, 1999; Yan et al., 2016; Zheng et al., 2018), with research interest then extending to the subsequent experience and following behaviour of the tourist as a result of the visit to the site (Cento & Angeli, 2020; Lacanienta et al., 2020; Oren et al., 2021; Sofia & Marques, 2020).

From sites to motivations

Later, in the 2000s scholars were requesting more attention towards the consumption, and the demand side of dark tourism, particularly tourists visiting dark sites and attractions (Dunkley, 2007; Dunkley et al., 2011, Hughes, 2008; Isaac et al., 2019; Sharpley & Stone, 2009; Stone 2006), simply asking: How tourists consume dark attractions? Thanks to these calls for attention, research has transformed and began to discover the demand of dark tourism and its nature, which proposed and theorized range of motives (Isaac & Çakmak, 2014; Isaac et al., 2019; Light, 2017; Tarlow, 2005; Wight & Lennon, 2002). In addition, studies of death and its relationship with consumption have gained meaningful consideration in the literature (Biran & Buda, 2018; Dobscha, 2016; Isaac et al., 2019; Podoshen et al., 2015b; Stone & Sharpley, 2008).

Despite the fact that many scholars tried to explain the tourist's motivation behind visiting dark tourism sites, which was based on assumptions and speculation and did now no longer have any empirical studies to help their theories (Ivanova & Light, 2018). The loss of research resulted in calls for more attention toward the tourists at dark sites (Huang & Hsu, 2009; Hughes, 2008; Seaton & Lennon, 2004; Stone, 2005). Therefore, in the mid-2000s the focus of the research has moved towards researching the nature of the demand for dark tourism attractions (Ivanova & Light, 2018). But still as Seaton and Lennon (2004, p. 82) state: '[T]here are extra questions than answers about visitor's motivations in dark tourism'. Nevertheless, a good range of motivations emerged within the literature of dark tourism are both empirically and conceptually addressed (Isaac & Çakmak, 2014, p. 168) (see Ashworth, 2002, 2004; Ashworth & Hartmann, 2005; Isaac & Çakmak, 2014; Isaac et al., 2019; Mangwane et al., 2019; Seaton & Lennon, 2004; Rojek, 1997; Seaton, 1999; Tarlow, 2005; Wight & Lennon, 2002) and 'most commonly by Dann (1998), that seeking experiences that challenge tourists their own sense of mortality – may be considered, one reason for participating in dark tourism' (Isaac & Çakmak, 2014, p. 168; Joly, 2010).

From an empirical perspective, recent data on tourist motivations of visiting dark sites have been examined from multiple perspectives, particularly motivational elements in potential tourists (Isaac & Çakmak, 2014; Isaac et al., 2019; Weaver et al., 2018). Furthermore, a number of studies have explored that individuals visit sites of dark tourism for personal reasons (Ashworth & Isaac, 2015), such as desire or opportunity for education, learning, understanding of what happened at the site (Best, 2007; Biran et al., 2011; Brida et al., 2013, 2016; Isaac & Çakmak, 2014, 2016; Kang et al., 2012; Sharpley, 2012; Thurnell-Read, 2009; Winter, 2011; Yan et al., 2016). Isaac and Çakmak (2014, p.172) examined the motivations of visitors to a former Second World War transit camp Westerbork (iconic dark site in the Netherlands) with 'self-understanding', 'curiosity', 'conscience', a 'must-see' site, and 'exclusiveness'. Other studies have found (Dunkley, 2007) different motives to visit 'dark' sites extend beyond a fascination with death, and this line of thought is in agreement with many scholars (e.g. Ashworth, 2002, 2004; Biran et al., 2011; Sharpley, 2005). The studies conducted by Biran et al. (2011); Kang et al. (2012); Rittichainuwat (2008); Zhang et al. (2016, p. 453) 'seeking to immerse themselves close to death' (Podoshen, 2013, p. 269), and Raine (2013)'s study in which he states visitors were interested in death and horror and were motivated by morbid curiosity. Korstanje and George (2015) claim that the primary reason for visiting a dark tourism is linked to the fascination with death. In brief, it is clear from the discussion on motivation that dark tourism continues to be eclectic and academically fragile, and ultimately understanding of the phenomenon of dark tourism remains incomplete (Buda, 2015a; Carrigan, 2014).

From motivations to experience

Stone (2018b, p. 510) states, 'fundamental to the study of dark tourism is the study of tourist experience at such sites. Henceforth, the need exists for more focused studies into tourist experience within dark tourism that reflect an ever-increasing diversity and complexity and the meaning and significance thereof'. Therefore, the tourist experience within dark tourism can only be understood by exploring specific contexts within which it occurs. Stone (2012b) and Sharpley (2005), in their work, established theoretical frameworks to know dark tourism experiences from a thanatological consumption perspective; supplementary empirical studies that classify shared dimensions in dark tourism are very scarce and tend to lack mechanism for further inspections (Nawijn & Fricke, 2015). Moreover, while significant contributions have been devoted to the diversity of typologies of dark tourism, less attention has been placed on their visitors (Biran et al., 2011) and what experience visitors gain.

Light (2017, p. 287) states, 'examination of visitors' experiences has been eclectic in its approach, focus and context (with most focus on the "darker" sites), but a recurring theme is that such experiences are complex and multi-layered and that, far from being a superficial encounter, a visit to a dark site has the potential to be profound and highly meaningful'. According to Tarlow (2005), 'heritage sites, which often are presented as a case study example for dark tourism (e.g., Auschwitz, Ground Zero), are spaces where tourists are involved in various, often not related, experiences' (Isaac & Çakmak, 2014, p. 175). Dark tourism sites evoke different experiences for different visitors and tourists. Therefore, there is a need for more empirical studies and discussions examining the nature of the tourist experience (Light, 2017), and the emotions they evoke. Dark experiences are common in tourists' experiences but have not yet received the scholarly attention they warrant. Nevertheless, experience of tourists vary and diverse, since visitors can engage with differing types of attractions (atrocities, prisons) through wide and alternative ways (Ashworth & Hartmann, 2005; Yankholmes & McKercher, 2015a, 2015b) counting on their motivations, their cultural background (Jamal & Lelo, 2011; Rittichainuwat, 2008; Zhang et al., 2016); the degree of their personal connection to the location (Ashworth, 2004; Ashworth & Hartmann, 2005; Cohen, 2011; Poria et al., 2004). That said, given the dynamic and sophisticated nature of dark tourism experiences, its global reach, and therefore, the ever-increasing significance given to dark tourism as attention on contemporary consumption, the 'experience economy' within dark tourism is still to be an important area for scholarly exploration (Stone, 2018b, p. 510).

From experience to emotions

'In a study of affect and emotion in dark tourism, a geographical perspective offers more food for thought regarding of disciplines used to investigate emotion and related constructs' (Martini & Buda, 2020; Volo, 2021, p. 4). This line of thought may advance the research into death and emotions that evoke during the visit. Therefore, it is suggested here that this approach in examining emotions at dark tourism sites is needed to understand the tourist experience and how these emotions evoke different experiences.

Ashworth and Isaac (2015) claim that, despite the variability of a classification of emotions, there is an inefficient prelude to answer important questions like 'What human emotions are evoked? Which of those are often classified as dark or light?' (p. 320). Emotions are a part of any sort of tourist experience, therefore, dark tourism is not a singular and single feature, by emotion themselves. However, Nawijn et al. (2016) proposed that the nature of these emotions is varying darker forms of dark tourism from other types of tourism. Dark tourism is considered emotionally laden (Nawijn & Fricke, 2015), and its experience is subjective and personal (Martini & Buda, 2018, 2020), together with both positive and negative experiences (Iliev, 2020; Sharma & Nayak, 2019). Positive emotions, for example (Best, 2007; Biran et al., 2014; Nawijn & Fricke, 2015; Thurnell-Read, 2009), and negative emotions (Austin, 2002; Miles, 2002; Nawijn & Biran, 2018) in dark tourism have been studied by a wide variety of academics. Cave and Buda (2018, pp. 707–726) called it the "emotional souvenirs process". Zhang et al. (2016) in their research recognized five affective experiences, including fear, sorrow, shock, appreciation, and depression, while Zheng et al. (2019) took a feature of dark tourism experiences as a diverse and undecided emotional.

Sadness, sorrow, grief, anger, and horror are emotions which were most commonly found in literature (Austin, 2002; Brown, 2016; Dunkley et al., 2011; Isaac & Çakmak, 2016; Kidron, 2013; Mowatt & Chancellor, 2011; Zhang et al., 2016). Other academics also reported anger (Mowatt & Chancellor, 2011), disappointment (Podoshen, 2013), shock and fear (Buda et al., 2014; Zheng et al., 2017), and sense of hope (Koleth, 2014). Moreover, Podoshen et al. (2015a) found in their research that visitors experienced repulsion and disgust. One could think that dark sites trigger only negative emotions in the visitors, but that is not exactly true. According to Nawijn et al. (2016), some dark sites can spark positive and also negative emotions. Nevertheless, the study by Oren et al. (2021) indicates that negative feelings motivate tourist to visit and clearly specifies that tourists' emotions that are classified in literature as negative play an important role in the tourist experience. With regard to positive ones, scholars stated a sense of hope and stimulation of pride (Cheal & Griffin, 2013; Koleth, 2014; Sharpley, 2012). However, Ashworth and Isaac (2015, p. 320) questioned the classification of negative and positive emotions when they stated that negative emotion can sometimes lead to a positive outcome and another way around. Emotional reactions to the tourism experience in dark tourism play a central role in determining tourists' satisfaction and behavioural intentions (Hosany & Prayag, 2013). The behavioural changes that are sought in dark tourism studies remain incomplete and mostly small scale (Ashworth & Isaac, 2015).

Based on the numerous studies, it can be concluded that 'visiting a dark site is a profoundly emotional experience' (Light, 2017, p. 288). Most of the time, this experience is characterized by a variety of emotions as the visit to a dark site is more than sightseeing or superficial voyeurism (Light, 2017, p. 288; Nawijn et al., 2015). Visit of dark site can even be a transformative power since a site offers the visitor with the opportunity for connection, engagement, and understanding (Cohen, 2011; Dunkley et al., 2011; Koleth, 2014; Light, 2017; Roberts & Stone, 2014).

To conclude, it is remarkable how important negative feelings are to the heritage experience (Oren et al., 2021). Site managers need to be aware of the factors and the processes that can trigger different visitors' emotions (Straker & Wrigley, 2015), as well as a dark tourism research has confirmed the importance of emotional engagement in dark tourism sites (Buda et al., 2014; Frew, 2018; Hede & Hall, 2012), because emotions affect the humans' daily lives, their activities, and subsequently their decision-making (Straker &Wrigley, 2015).

Structure of the book

The present edited volume advances conceptual as well as empirical research on dark tourism. The collection of chapters in this volume offers original and advanced international tourism research studies that are entrenched within interdisciplinary and transdisciplinary conceptual and methodological thought in the study of dark tourism. This edited volume comprises a series of chapters that capture a glimpse of the types of cutting-edge thinking and academic research in the domain of dark tourism studies. The Editor hopes that by reading this edited volume academics and practitioners will be encouraged to activate their own studies that will eloquently advance this body of investigation.

Chapter one by Kennell and Powell, 'Dark tourism and World Heritage Sites: a Delphi study of stakeholder perceptions of the development of dark tourism products', states that despite the growth of UNESCO World Heritage Site designations, little research has considered the relationship between dark tourism and World Heritage Sites. The aim of their research was to evaluate stakeholder's perceptions of the potential development of dark tourism to the Greenwich Maritime World Heritage Site in London, employing a qualitative Delphi Panel method. Their findings show that stakeholders are broadly supportive of tourism to the site and are positive about future tourism growth. However, they did not support the development of dark tourism to the site because it was perceived as inauthentic, messy, and sensationalist. They argue that future attempts to develop dark tourism at World Heritage Sites should include enhancing the knowledge of stakeholders about dark tourism.

Chapter two focuses on the significant gap in relation to the debate of crime-based dark tourism activities and its engagement with gender. The chapter presents a conceptual discussion on tourism to sites of female criminal activity, drawing parallels to comparable male crime locations. The scrutiny of online advertising for murder walking tours in the UK reveals gendered power dynamics wherein traditional, western gender roles are enforced through the removal of agency from women who engage in more violent crimes while simultaneously fetishizing women as victims of violence, especially sexual. The authors Adie and Snell argue that this is evident in the absence of female serial killers within organized dark tours, which often focus specifically on this sexual violence. Accordingly, the tourist activities that revolve around dark heritage sites, especially those that deal with violent criminal activity, support gendered stereotypes around 'acceptable' transgression.

MacCarthy and Rigne in Chapter three examined visitor's experience at the National Anzac Centre in Western Australia employing multiple qualitative methods. Their findings advise a significant visitor-thirst for the positive aspects of commemoration. They argue if a certain fascination with, and commodification of death defines popular dark tourism then commemorative tourism's relegation of death indicates exception. It would seem commemorists relegate death and darkness to mere context, while gravitas, and ritual and cultural validation transcend the superficial and the kitsch.

Chapter four by Sigala and Sotiropoulos's study contributes to the field by adopting a reflective autoethnographic approach for providing a better understanding of dark tourism experiences. Their study critically reflects on the researcher's immersive experiences at three USA dark sites (Ground Zero, Gettysburg, and Ellis Island). The study reveals that emotional engagement plays an important role in probing and helping visitors to generate meaning through their dark tourism experiences. Based on their findings, they proposed an adapted dark tourism typology framework whereby emotional engagement is used as an explanatory theoretical concept to better identify and understand the nuanced types of dark tourism experiences.

Chapter five by Wyatt, Leask, and Barron concentrates on designing dark tourism experiences at light dark attractions. Research appears under-developed regarding the influences on the design of interpretation at dark visitor attractions, particularly those considered lighter due to their edutainment agenda. The aim of their chapter was to critically discover the influences on the design of edutainment interpretation at three lighter dark visitor attractions, which are introduced as new attractions for study within dark tourism research. The findings of their study not only contribute to the study's conclusions and recommendations for future research in the realms of dark tourism and interpretation, but also contribute to enhancing interpretation design understanding for both dark tourism research and practice.

Chapter six written by Grinfelde and Veliverronena addresses the role of students' field trips to dark tourism sites in higher education in Latvia. This chapter studies the importance of the educational experience of the students at dark tourism sites, with a particular focus on the effects after their exposure to dark sites – inner-directed and other behavioural activities. With inner-directed post-visit effects, they understand behavioural activities guided by internal values, e.g. reading and researching more about the topic, rethinking students' personal values, reconsidering individual and collective responsibility, and rethinking the past and future. With others directed behavioural effects, they understand activities that involve interaction with other people, such as talking with peers, telling about the site to friends or family, planning to revisit a site with friends. The study is based on longitudinal data collected after field trips to dark tourism sites in Latvia (2014–2019). They argue that students appreciate the educational value of the dark tourism sites, and better-designed sites induce more post-visit effects. The final chapter (Chapter 7) written by Jang, Sakamoto, and Funck focuses on dark tourism as educational tourism, through 'hope tourism' in Fukushima, Japan. They examined how dark tourism has been adopted and developed in Japan and advocates how it can be linked to educational tourism, through the case of a 'Hope Tourism Guided Tour' held in Fukushima in 2018. They claim that in Japan, dark tourism has developed into a new form, nested within educational tourism.

References

Ashworth, G.J. (2002). Review of Dark tourism: The attraction of death and disaster by J. Lennon and M. Foley. *Tourism Management*, 23(2), 187–193.

Ashworth, G.J. (2004). Tourism and the heritage of atrocity: Managing the heritage of South African apartheid for entertainment. In T. V. Singh (Ed.), *New horizons in tourism: Strange experiences and stranger practices* (pp. 95–108). Wallingford: CAB International.

Ashworth, G., & Hartmann, R. (2005). Introduction: Managing atrocity for tourism. In G. Ashworth, & R. Hartmann (Eds.), *Horror and human tragedy revisited: The management of sites of atrocities for tourism* (pp. 1–14). New York: Cognizant Communication Corporation.

Ashworth, G.J., & Isaac, R.K. (2015). Have we illuminated the dark? Shifting perspectives on 'dark' tourism. *Tourism Recreation Research*, 40(3), 316–325.

Austin, N. (2002). Managing heritage attractions: Marketing challenges at sensitive historical sites. *International Journal of Tourism Research*, 4(6), 447–457.

Best, M. (2007). Norfolk islands: Thanatourism, history and visitor emotions. *Shima: The International Journal of Research into Island Cultures*, 1(2), 30–48.

Biran, A., Liu, W., Li, G., & Eichhorn, V. (2014). Consuming post-disaster destinations: The case of Sichuan, China. *Annals of Tourism Research*, 47, 1–17.

Biran, A., & Buda, D.M. (2018) Unravelling fear of death motives in dark tourism. In P. Stone, R. Hartmann, A. Seaton, R. Sharpley & L. White (Eds.), *The Palgrave handbook of dark tourism studies*. (pp. 515–532). London: Macmillan.

Biran, A., Poria, Y., & Oren, G. (2011). Sought experiences at (dark) heritage sites. *Annals of Tourism Research*, 38(3), 820–841.

Brida, J.G., Disegna, M., & Scuderi, R. (2013). Visitors of two types of museums: A segmentation study. *Expert Systems with Applications*, 40(6), 2224–2232.

Brida, J.G., Nogare, C.D., & Scuderi, R. (2016). Frequency of museum attendance: Motivation matters. *Journal of Cultural Economics*, 4(3), 261–283.

Brown, L. (2016). Tourism and pilgrimage: Paying homage to literary heroes. *International Journal of Tourism Research*, 18(2), 167–175.

Buda, D.M. (2005). *Affective tourism: Dark routes in conflict. Tourism Geographies*, 23(5–6), 963–984. New York: Routledge.

Buda, D.M., d'Hauteserre, A.-M., & Johnston, L. (2014). Feeling and tourism studies. *Annals of Tourism Research*, 46, 102–114.

Carrigan, A. (2014). Dark tourism and postcolonial studies: Critical intersections. *Postcolonial Studies*, 17(3), 236–250.

Cave, J., & Buda, D. (2018). Souvenirs in dark tourism: Emotions and symbols. In R. Stone, R. Hartmann, T. Seaton, R. Sharpley, & L. White (Eds.), *The Palgrave handbook of dark tourism studies* (pp. 707–726). Basingstoke: Palgrave Macmillan.

Cento, A., & Amgeli, D. (2020) Emotions and critical thinking at a dark heritage site: Investigating visitor's reactions to a first world was museum in Slovenia. *Journal of Heritgae Tourism*. http://dx.doi.org/10.1080/17438 73X.2020.1804918

Cheal, F., & Griffin, T. (2013). Pilgrims and patriots: Australian tourist experiences at Gallipoli. *International Journal of Culture, Tourism and Hospitality Research*, 7(3), 227–241.

Cohen, E.H. (2011). Educational dark tourism at an in Populo site: The Holocaust Museum in Jerusalem. *Annals of Tourism Research*, 38(1), 193–209.

Dann, G.M.S. (1998). *The dark side of tourism, Etudes et Rapports Serie L, Sociology/ Psychology/ Philosophy/Anthropology (Vol 14)*. Aix-en-Provence: Centre International de Recherches et d'Etudes Touristiques.

Dobscha, S. (2016). Death in a consumer culture. Oxford: Routledge.

Dunkley, R.A. (2007). Re-peopling tourism: A 'hot approach' to studying thanatourist experiences. In I. Ateljevic, A. Pritchard & N. Morgan (Eds.), *The critical turn in tourism studies: Innovative research methodologies* (pp. 371–385). Amsterdam: Elsevier.

Dunkley, R., Morgan, N., & Westwood, S. (2011). Visiting the trenches: Exploring meanings and motivations in battlefield tourism. *Tourism Management*, 32(4), 860–868.

Foley, M., & Lennon, J.J. (1996). JFK and dark tourism: A fascination with assassination. *International Journal of Heritage Studies*, 2(4), 198–211.

Frew, E. (2018). Exhibiting death and disaster: Museological perspectives. In P. Stone, R. Hartmann, T. Seaton, R. Sharpley & L. White (Eds.), *The Palgrave handbook of dark tourism studies* (pp. 693–706). Basingstoke: Palgrave Macmillan Publishers Ltd.

Hartmann, R. (2014). Dark tourism, thanatourism and dissonance in heritage tourism management. New directions in contemporary tourism research. *Journal of Heritage Tourism*, 9(2), 166–182.

Hede, A.-M., & Hall, J. (2012). Evoked emotions: Textual analysis within the context of pilgrimage tourism to Gallipoli. Advances in culture. *Tourism and Hospitality Research*, 6, 45–60.

Hosany, S., & Prayag, G. (2013). Patterns of tourists' emotional responses, satisfaction, and intention to recommend. *Journal of Business Research*, 66(6), 730–737.

Huang, S., & Hsu, C.H.C. (2009). Travel motivation: Linking theory to practice. *International Journal of Culture, Tourism and Hospitality Research*, 3(4), 287–295.

Hughes, R. (2008). Dutiful tourism: Encountering the Cambodian genocide. *Asian- Pacific Viewpoint*, 49(3), 318–330.

Iliev, D. (2020). Consumption, motivation and experience in dark tourism: A conceptual and critical analysis. *Tourism Geographies*, 23(5–6), 963–984.

Isaac, R.K., & Çakmak, E. (2014). Understanding visitor's motivations at sites of death and disaster: The case of former transit camp Westerbork, The Netherlands. *Current Issues in Tourism*, 17(2), 164–179.

Isaac, R., & Çakmak, E. (2016). Understanding the motivations and emotions of visitors at Tuol Sleng genocide prison museum (S-21) in Phnom Penh, Cambodia. *International Journal of Tourism Cities*, 2(3), 232–247.

Isaac, R.K., Nawijn, J., van Liempt, A. & Gridnevskiy, K. (2019) Understanding Dutch visitors' motivations to concentration camp memorials. *Current Issues in Tourism*, 22(7), 747–762.

Ivanova, P., & Light, D. (2018). 'It's not that we like death or anything': Exploring the motivations and experiences of visitors to a lighter dark tourism attraction. *Journal of Heritage Tourism*, 13(4), 356–369. http://dx.doi.org/10.1080/1743873x.2017.1371181

Jamal, T., & Lelo, L. (2011). Exploring the conceptual and analytical framing of dark tourism: From darkness to intentionality. In R. Sharpley & P. Stone (Eds.), *Tourist experience: Contemporary perspectives* (pp. 29–42). Abingdon: Routledge.

Joly, D. (2010). *The dark tourist: Sightseeing in the world's most unlikely holiday destinations*. London: Simon and Shuster.

Kang, E.J., Scott, N., Lee, T.J., & Ballantyne, R. (2012). Benefits of visiting a 'dark tourism' site: The case of the Jeju April 3rd Peace Park, Korea. *Tourism Management*, 33(2), 257–265.

Kidron, C. (2013). Being there together: Dark family tourism and the emotive experience of co-presence in the Holocaust past. *Annals of Tourism Research*, 41, 175–194.

Koleth, M. (2014). Hope in the dark: Geographies of volunteer and dark tourism in Cambodia. *Cultural Geographies*, 21(4), 681–694.

Korstanje, M., & George, B. (2015). Dark tourism: Revisiting some philosophical issues. *e-Review of Tourism Research*, 12(1/2), 127–136.

Lacanienta, A., Ellis, G. Hill, B., Freeman, P., & Jiang, J. (2020) Provocation and related subjective experiences along the dark tourism spectrum. *Journal of Heritage Tourism*, 15(6), 626–647.

Lennon, J. (2010). Dark tourism and sites of crime. In D. Botterill & T. Jones (Eds.), *Tourism and crime: Key themes* (pp. 215–228). Oxford: Goodfellow Publishers.

Light, D. (2017). Progress in dark tourism and thanatourism research: An uneasy relationship with heritage tourism. *Tourism Management*, 61, 275–301. http://dx.doi.org/10.1016/j.tourman.2017.01.011

Mangwane, J., Hermann, U.P., & Lenhard, A.I. (2019). Who visits the apartheid museum and why? An exploratory study of the motivations to visit a dark tourism site in South Africa. *International Journal of Culture, Tourism and Hospitality Research*, 13(3), 273–287.

Miles, W. (2002). Auschwitz: Museum interpretation and darker tourism. *Annals of Tourism Research*, 29(4), 1175–1178.

Martini, A., & Buda, D.M. (2019). Analysing affects and emotions in tourist e-mail interviews: A case in post-disaster Tohoku, Japan. *Current Issues in Tourism*, 1–12. 22(19), 2353–2364.

Martini, A., & Buda, D.M. (2020). **Dark tourism and affect: Framing places of death and disaster.** *Current Issues in Tourism*, 23(6), 679–692.

Mowatt, R.A., & Chancellor, C.H. (2011). Visiting death and life: Dark tourism and slave castles. *Annals of Tourism Research*, 38(4), 1410–1434.

Nawijn, J., & Biran, A. (2018). Negative emotions in tourism: A meaningful analysis. *Current Issues in Tourism*, 22(19), 2386–2398.

Nawijn, J., & Fricke, M.-C. (2015). Visitor emotions and behavioural intentions: The case of Concentration Camp Memorial Neuengamme. *International Journal of Tourism Research*, 17(3), 221–228. http://dx.doi.org/10.1002/jtr.1977

Nawijn, J., Isaac, R.K., Gridnevskiy, K., & van Liempt, A. (2015). Holocaust concentration camp memorial sites: An exploratory study into expected emotional response. *Current Issues in Tourism*, 21(2), 175–190.

Nawijn, J., Isaac, R.K., van Liempt, A., & Gridnevskiy, K. (2016). Emotion clusters for concentration camp memorials. *Annals of Tourism Research*, 61, 244–247.

Oren, G., Shani, A., & Poria, Y. (2021) Dialectical emotions in a dark heritage site: A study at the Auschwitz Death Camp. *Tourism Management*, 82, 104194 https://doi.org/10.1016/j.tourman.2020.104194

Podoshen, J.S. (2013). Dark tourism motivations: Simulation, emotional contagion and topographic comparison. *Tourism Management*, 35, 269–289.

Podoshen, J.S., Andrzejewski, S.A., Venkatesh, V., & Wallin, J. (2015a). New approaches to dark tourism inquiry: A response to Isaac. *Tourism Management*, 51, 331–334.

Podoshen, J.S., Venkatesh, V., Wallin, J., Andrzejewski, S.A., & Jin, Z. (2015b). Dystopian dark tourism: An exploratory examination. *Tourism Management*, 51, 316–328.

Poria, Y., Butler, R., & Airey, D. (2004). The core of heritage tourism: Distinguishing heritage tourists from tourists in heritage places. *Annals of Tourism Research*, 30, 238–254.

Raine, R. (2013). A dark tourism spectrum. *International Journal of Culture, Tourism and Hospitality Research*, 7, 242–256.

Rittichainuwat, B. N. (2008). Responding to disaster: Thai and Scandinavian tourists' motivation to visit Phuket, Thailand. *Journal of Travel Research*, 46, 422–432.

Roberts, C., & Stone, P. (2014). Dark tourism and dark heritage: Emergent themes, issues and consequences. In I. Convery, G. Corsane, & P. Davies (Eds.), *Displaced heritage: Responses to disaster, trauma and loss* (pp. 9–18). Woodbridge: Boydell Press.

Rojek, C. (1997) Indexing, dragging and the social construction of tourist sights. In C. Rojek & J. Urry (Eds.), *Tourism, cultures and transformations of travel and theory* (pp. 52–74). London: Routledge.

Seaton, A (1999) War and thanatourism: Waterloo 1815–1914. *Annals of Tourism Research*, 26(1), 130–159.

Seaton, A.V. (1996). Guided by the dark: From thantopsis to thanatourism. *International Journal of Heritage Studies*, 2(4), 234–244.

Seaton, A.V., & Lennon, J.J. (2004). Thanatourism in the early 21st Century: Moral panic, ulterior motives and alterior desires. In T.V. Singh (Ed.), *New horizons in tourism: Strange experiences and stranger practices* (pp. 63–82). Wallingford: CAB International.

Sharma, P., & Nayak, J. (2019). Examining experience quality as the determinant of tourist behavior in niche tourism: An analytical approach. *Journal of Heritage Tourism*, 15(1), 76–92.

Sharpley, R. (2005). Travels to the edge of darkness: Towards a typology of dark tourism. In C. Ryan, S. Page, & M. Aitken (Eds.), *Taking tourism to the limits: Issues, concepts and managerial perspectives* (pp. 217–228). Oxford: Elsevier.

Sharpley, R. (2012). Towards an understanding of 'genocide tourism': An analysis of visitor's accounts of their experience of recent genocide sites. In R. Sharpley & P.R. Stone (Eds.), *Contemporary tourist experience: Concepts and consequences* (pp. 95–109). London: Routledge.

Sharpley, R., & Stone, P. R. (2009). Life, death and dark tourism: Future research directions and concluding comments. In R. Sharpley & P.R. Stone (Eds.), *The darker side of travel: The theory and practice of dark tourism* (pp. 247–251). Bristol: Channel View.

Sofia, M. & Marques, L. (2020) Dystopian dark tourism: affective experiences in Dismaland. Tourism Geograohies DOI: 10.1080/14616688.2020.1795710

Stone, P.R. (2005). Dark tourism consumption e a call for research. *e-Review of Tourism Research*, 3(5), 109–117.

Stone, P. (2006). A dark tourism spectrum: Towards a typology of death and macabre related tourist sites, attractions and exhibitions. Tourism, 52, 145-160.

Stone, P. R. (2012b). Dark tourism as 'mortality capital': The case of Ground Zero and the significant other dead. In R. Sharpley, & P. R. Stone (Eds.), Contemporary tourist experience: Concepts and consequences (pp. 71-94). London: Routledge.

Stone, P., & Sharpley, R. (2018) (eds) *The Palgrave handbook of dark tourism studies*. London: Palgrave.

Stone, P.R (2018b) The dark tourist experiece. In: R. Stone P., Hartmann R., Seaton T., Sharpley R., White L. (eds) The Palgrave Handbook of Dark Tourism Studies (pp. 509-513). Palgrave Macmillan, London.

Straker, K., & Wrigley, C. (2015). The role of emotion in product, service and business model design. *Journal of Entrepreneurship, Management and Innovation*, 11(1), 11–28.

Tarlow, P.E. (2005). Dark tourism: The appealing 'dark' side of tourism and more. In M. Novelli (Ed.), *Niche tourism: Contemporary issues, trends and cases* (pp. 47–58). Amsterdam: Elsevier.

Thurnell-Read, T. (2009). Engaging Auschwitz: An analysis of young travellers' experiences of Holocaust tourism. *Journal of Tourism Consumption and Practice*, 1(1), 26–52.

Volo, S. (2021). The experience of emotion: Directions for tourism design. *Annals of Tourism Research*, 86, 103097.

Weaver, D., Tang, C., Shi, W., Huang, M. F., Burns, K., & Sheng, A. (2018). Dark tourism, emotions, and post-experience visitor effects in a sensitive geopolitical context: A Chinese case study. *Journal of Travel Research*, 56, 1–15.

Wight, A.C., & Lennon, J.J. (2002). Towards an understanding of visitor perceptions of 'dark' attractions: The case of the Imperial war museum of North Manchester. *Journal of Hospitality and Tourism*, 2(2), 105–122.

Winter, C. (2011). First World War cemeteries: Insights from visitor books. *Tourism Geographies*, 13(3), 462–479.

Yan, B.-J., Zhang, J., Zhang, H.-L., Lu, S.-J., & Guo, Y.-R. (2016). Investigating the motivation-experience relationship in a dark tourism space: A case study of the Beichuan earthquake relics, China. *Tourism Management*, 53, 108–121.

Yankholmes, A., & McKercher, B. (2015a). Rethinking slavery heritage tourism. *Journal of Heritage Tourism*, 10(3), 233–247.

Yankholmes, A., & McKercher, B. (2015b). Understanding visitors to slavery heritage in Ghana. *Tourism Management*, 51(12), 22–32.

Zhang, H., Yang, Y., Zheng, C., & Zhang, J. (2016). Too dark to revisit? The role of past experiences and intrapersonal constraints. *Tourism Management*, 54, 452–464.

Zheng, C., Zhang, J., Zhang, H., & Qian, L. (2017). Exploring sub-dimensions of intrapersonal constraints to visiting "dark tourism" sites: A comparison of participants and non-participants. *Asia Pacific Journal of Tourism Research*, 22(1), 21–33.

Zheng, C. Zhang, J., Qian, L., Jurowski, C., Zhang, H., & Yan, B. (2018). The inner struggle of visiting 'dark tourism' sites: Examining the relationship between perceived constraints and motivations. *Current Issues in Tourism*, 21(15), 1710–1727.

Zheng, C., Zhang, J., Qiu, M., Guo, Y., & Zhang, H. (2019). From mixed emotional experience to spiritual meaning: Learning in dark tourism places. *Tourism Geographies*, 22(1), 105–126.

Dark tourism and World Heritage Sites: a Delphi study of stakeholder perceptions of the development of dark tourism products

James Kennell ⓘ and Raymond Powell

ABSTRACT

Dark tourism has attracted increasing academic attention, but the extent to which it exists as a separate form of tourism from heritage tourism is not yet clear. Despite the growth of UNESCO World Heritage Site designations, little research has considered the relationship between dark tourism and World Heritage Sites. Because the development of dark tourism is beset with ethical concerns, heritage professionals can have negative perceptions about the acceptability or attractiveness of it for the sites that they are involved in managing. This research used a qualitative Delphi Panel method to evaluate stakeholder perceptions of the potential development of dark tourism to the Greenwich Maritime World Heritage Site in London, United Kingdom. The findings show that stakeholders are broadly supportive of tourism to the site and positive about future tourism growth. Despite this, they did not support the development of dark tourism to the site because it was perceived as inauthentic, tacky and sensationalist. In order to address this issue, recommendations are made that future attempts to develop dark tourism at WHS should involve enhancing the knowledge of stakeholders about dark tourism, and of the resources within their sites that could be included in a dark tourism offer to tourists.

Introduction

Despite twenty years of academic research into the nature and practices of dark tourism (Ashworth & Isaac, 2015), and its emergence in popular culture (e.g. Fryer, 2018), the extent to which it exists as a separate form of tourism is still contested (Light, 2017). It is not clear whether visitation to sites and attractions associated with death, disaster and suffering is something new, or simply a reframing of well-established heritage offerings by a ghoulish and attention-hungry media (Roberts, 2018). Because of this, dark tourism can suffer from negative perceptions from heritage tourism professionals, who worry about the dangers of promoting their products in a way that attracts sensationalist attention from tourists. The aim of this research was to assess stakeholder perceptions of dark tourism at a World Heritage Site (WHS), with a view to assessing its acceptability and viability in an established heritage setting.

To this end, this study critically examined these tensions through a case study of the Maritime Greenwich UNESCO World Heritage site in London, United Kingdom: a tourist destination that receives in excess of nineteen million visitors every year, including over one million overnight stays (Visit Greenwich, 2018). The site is diverse, containing military, religious, scientific, natural and architectural heritage, as well as being home to two universities and a national museum

(Maritime Greenwich World Heritage Site, 2019). This research applies the Delphi study technique (Lin & Song, 2015) to survey the views of these stakeholders on the acceptability of dark tourism, and ways in which this could develop in the future. Over three rounds of questions, a panel of stakeholders were asked to define, explore and forecast the role of dark tourism on the site.

Literature review

Dark tourism in context

Despite the years of research into dark tourism (Dale & Robinson, 2011; Lennon & Foley, 2000; Stone, 2013; Tarlow, 2005) and the growth of the dark tourism market (Biran & Hyde, 2013; Biran et al., 2011; Stone & Sharpley, 2008), and the academic interest in this field (Ashworth & Isaac, 2015) there has been little interest shown in understanding the relationship between dark tourism and other forms of tourism. As Light (2017, p. 275) notes: ' … two decades of research have not convincingly demonstrated that dark tourism and thanatourism are distinct forms of tourism, and in many ways they appear to be little different from heritage tourism.'

Dark tourism is frequently described as having a spectrum of darkness (Miles, 2002; Stone, 2006). At the darkest end of the spectrum, sites are categorised largely on the basis of recent and actual suffering and death. There is usually an educational and commemorative rationale which underpins these sites, which very often is the authentic place of suffering. At the lighter end of this spectrum, tourism products are associated with 'fright tourism' (Bristow, 2020). This includes haunted houses, ghost tours, and scary stories, for example, with an entertainment, rather than educational or commemorative rationale. As Bucior (2020) has shown in a study of the interpretation of the Gettysburg battlesite in the USA, however, these two poles are not mutually exclusive. Interpretive tools such as ghost tours can provide alternatives to the 'Authorized Heritage Discourse' of dark sites, especially where these involve contested narratives. The notion that the diversity of sites in dark tourism can be categorised using a straightforward scale has been substantially critiqued (Ashworth & Isaac, 2015; Dale & Robinson, 2011; Ivanova & Light, 2018); most frequently this has been due to subjective classifications of sites and histories as 'dark,' but the persistence of the spectrum in the literature indicates its enduring utility as a descriptive tool.

Dark tourism and heritage tourism

Whilst a subset of heritage tourism may be considered dark, there is no readily distinguishable divide between aspects of heritage tourism and aspects of dark tourism *per se*. Richards (2001) says it is necessary to broaden the categorisation of heritage attractions to include intangible ideas, including those that relate to the ideas such as statehood, history and struggle, often typical of dark sites (Murtagh et al., 2017). There is a clear link between heritage tourism and dark tourism, theoretically inextricable in the majority of cases (Hartmann, 2014), and the term 'dark heritage' (Kamber et al., 2016) is already in use to capture these associations.

Biran et al. (2011) identify that the experience of visitors is important to the conceptualisation of dark tourism, which is largely the product of on-site interpretation, and it is unlikely to be simply a fascination with death which encourages visitation to dark sites. It is not necessary to have a morbid interest in death to be fascinated by aspects of death, especially when such narratives have personal or national significance. For example, ' … Australians and New Zealanders visiting Gallipoli are enaged in a profound heritage experience and are not interested in death itself' (p. 822). Other motivations for visiting dark sites, such as a desire for novelty, nostalgia, curiosity, entertainment and pilgrimage are the same as motivations for visiting heritage sites (Ashworth, 2004; Biran et al., 2011; Hyde & Harman, 2011; Stone & Sharpley, 2008; Tarlow, 2005). The motivation of pilgrimage is particularly relevant for many dark tourism sites, when expanded to include more secular definitions of the term, where religious motivations are less important than other commemorative aspects of visitation and the sense that these are sites which 'add meaning to life' (Collins-Kreiner, 2016, p. 1187).

Postmodern contexts for the growth of dark tourism (Powell & Kennell, 2016) offer competing conceptualisations of this, in the context of increasing interest in utopian and dystopian visions of the world (Farkic, 2020; Podoshen et al., 2015). There is no agreement in the literature about the categorisations of dark tourism motivations, and all that can be certain is that there are a wide variety of these (Isaac & Çakmak, 2014; Raine, 2013).

The dark tourism literature demonstrates the current inadequacies and impreciseness of definitions of dark tourism to date. Dark sites are perceived as being associated with death, disaster and frequently genocide, yet many dark sites are also dynamic and elevating, structured servicescapes (Magee, 2018). Thus it would be wrong to assume that dark tourism is only concerned with the macabre. Undoubtedly that remains a fascination, but as the development of dark attractions is a relatively new phenomenon (Sharpley, 2005), notwithstanding the long-established practice of travelling to sites of suffering since medieval times, sites at the darkest end of the spectrum are only a fraction of sites which record the history of humanity and as dark histories are intermingled inextricably with all aspects of human history, the significance of dark tourism motivations may be overestimated by researchers.

There can be squeemishness around the commercial exploitation of dark sites. Dann (1994) identifies that there is the potential to capitalise on the 'product of dark tourism' and 'milk the macabre' (p. 61). Ethical considerations abound when establishing attractions with dark themes (Stone & Sharpley, 2008). WHS need to consider the authenticity of new tourism offers, to avoid accusations of commercialisation and 'cashing in,' which may complicate any desire to increase revenue. Dark and heritage attractions cover a wide spectrum of authenticity, and the perceived authenticity of a site is important. Attractions at the lightest end of the darkness spectrum (Stone, 2006) do, however, attract visitation without being authentic.

Dark tourism sites and attractions are often significant in the forming of national stories and identity and an understanding of often contested history (de-Miguel-Molina & Barrera-Gabaldón, 2019; Kennell et al., 2018; Lemelin et al., 2013) and the 'history wars' that can take place over such sites (Boyle, 2019). For WHS, there is a tension between the notion of 'universal' value, which is a condition of WHS status, and such 'contested' heritage, which implies the absence of a settled interpretation of a site (Rakic & Chambers, 2008).

Tourism to world heritage sites

In order to protect and preserve cultural and natural heritage from a range of threats, the United Nations Educational, Scientific and Cultural Organisation (UNESCO) instituted the World Heritage Convention in 1972 (Leask & Fyall, 2001) and have since designated 1121 properties as WHS, of which 869 are cultural, 213 are natural and 39 are mixed in nature (UNESCO, 2020). The official designation of a heritage site as of particular value can often lead to a rise in tourism (Dans & González, 2019; Kwiatek-Sołtys & Bajgier-Kowalska, 2019), and the award of WHS status can be transformative for many destinations in this respect (Cassel & Pashkevich, 2014), notwithstanding more broad critiques about the overall benefits of WHS inscription for tourism growth, on which evidence is mixed (Gao & Su, 2019; Mariani & Guizzardi, 2020).

Concerns about the authenticity of tourism offers at heritage sites feature prominently in the literature (Katahenggam, 2019; Nuryanti, 1996; Yi et al., 2018). A distinction should be drawn between the tangible heritage of the site and its value as assigned by its custodians, and tourist and resident perceptions of authenticity (Yi et al., 2018), as the tension between these two poles is often at the root of conflicts about tourism at WHS (Kim et al., 2018). As Dans and González (2019) have argued, the social value of heritage sites should be taken into account along with their economic, aesthetic and other values. Imon (2017) highlights that this can be a particularly pertinent issue for heritage sites within urban settings, where competing social and cultural values co-exist, necessitating integrated tourism planning and development activities in order to ensure that tourism to WHS is sustainable. In the case of China, Gao and Su (2019) found that WHS inscription functioned more effectively to preserve sites rather than to promote them as tourism destinations, showing that not all sites approach this dilemma from the same position, with some choosing not to invite the management challenges that come with increased visitation.

Tangible heritage resources are not a sufficient precondition for the development of tourism to WHS, it is still the case that other conditions must be met to develop a tourism destination using the WHS, including successful marketing campaigns (de Fauconberg et al., 2018), the inclusion of the resources within creative, dynamic experiences that attract tourists (Cassel & Pashkevich, 2014) and effective governance, including stakeholder management (Landorf, 2009; Su et al., 2017). Governance arrangements for WHS show considerable national variation, as it is the responsibility of states, and not UNESCO itself, to manage WHS (Adie & Amore, 2020; Su & Wall, 2012). A core function of WHS governance is the management of the interests and activities of diverse sets of stakeholders (Evans, 2002), which can include representatives of governments, cultural and heritage bodies, business, local communities, and users of the sites (Davey & Gillespie, 2014). Successful stakeholder engagement has been identified consistently as a prerequisite for sustainable tourism development in a variety of types of heritage contexts, and national settings (Liburd & Becken, 2017; Li et al., 2020; Rasoolimanesh & Jaafar, 2017; Timothy & Boyd, 2003). This research examines the potential for the development of dark tourism in a WHS destination, a type of tourism that can be accused of commercialising and trivialising more 'serious' heritage (Stone & Sharpley, 2008) and of 'milking the macabre' (Dann, 1994). Because of these concerns, it is vital to engage stakeholders at the early stage of potential product development, in order to evaluate the sustainability of any future tourism growth in this area.

Methodology

To this end, this study critically examined these tensions through a case study of the Maritime Greenwich UNESCO World Heritage site in London, United Kingdom: a tourist destination that receives in excess of nineteen million visitors every year, including over one million overnight stays (Visit Greenwich, 2018). For the purposes of this research, the Maritime Greenwich UNESCO World Heritage Site in London, United Kingdom was chosen to carry out research into stakeholder perceptions of the potential product development area of dark tourism. There were two reasons why this location as chosen. Firstly, the site is very diverse and contains numerous potential resources for a dark tourism product to draw upon. These include the National Maritime Museum for the United Kingdom, which contains numerous artefacts, archives and artworks linked to the history of naval warfare and the colonial expansion of the British Empire; The Old Royal Naval College, which was the site of a historic Royal palace where King Henry VIII and Elizabeth I were born, before becoming the training academy for the Royal Navy, as well containing the chapel where Admiral Lord Nelson lay in state after his death at the Battle of Trafalgar; St Alfege's church, the site of the martyrdom of an early English Christian saint; as well as numerous memorials, statues and collections and burial spaces linked to themes from across the dark tourism spectrum (Miles, 2002; Stone, 2006), some of which are interpreted through guided and self-guided tours (Maritime Greenwich World Heritage Site, 2019).

Secondly, as an urban WHS, this location presents an opportunity to survey the view of a wide range of stakeholders, who are likely to have competing priorities because urban heritage site are subject to multiple uses by users with often competing priorities (Imon, 2017). This research applied the Delphi study technique (Lin & Song, 2015) to survey the views of these stakeholders on the acceptability of dark tourism, and ways in which this could develop in the future. Over three rounds of questions, a panel of stakeholders were asked to define, explore and forecast the role of dark tourism on the site. Because of the diverse nature of the site, and the broad range of stakeholders involved in its operations and governance, this site provides a setting from which conclusions can be drawn that may be useful for future research and product development in other WHS contexts.

The Delphi method is a forecasting technique that has been in use since the 1950s when it was developed by researchers at the RAND Corporation (Habibi et al., 2014). It was informed by the pragmatist approach to knowledge, which bridges the interpretivist and post-positivist paradigms to provide practical guidance in decision making in complex scenarios (Brady, 2015; Day & Bobeva, 2005). It makes use of anonymous, expert opinion from a panel to consider the options for dealing with a complex problem, in order to build knowledge from consensus positions, which can avoid the problems caused by power-dynamics involved in face-to-face situations (Habibi et al., 2014). Avella (2016) explains that the Delphi method can be particularly advantageous when researching issues that are multi-disciplinary, involving lots of uncertainty and where anonymity is beneficial. Given the nature of dark tourism, it is likely that the research process will require participants to engage with complex issues with psychological, personal and social dimensions, suggesting that a qualitative Delphi technique may provide a forum for panellists to respond anonymously, and in depth, without fear of judgement from other participants.

The Delphi method has been applied in hugely diverse contexts meaning that it has been criticised for its methodological heterogeneity (Day & Bobeva, 2005; Habibi et al., 2014), but there are accepted general principles for qualitative Delphi designs, which are: purposive sampling; emergent design; anonymous and structured communication between participants and; thematic analysis (Brady, 2015).

This research utilised a conventional design, using multiple iterations to eventually allow for a consensus position to evolve from the panel, where the responses of the panel to each round of questions are analysed, summarised, and reflected back to the panel, along with a set of follow-up question, in a series of rounds. This process of asking for, analysing and reflecting is a reflection of the interpretivist approach that underpins the Delphi technique (Avella, 2016; Sobaih et al., 2012).

This research follows the process set out by Donohoe and Needham (2009), who reviewed the application of Delphi techniques in tourism and suggested a process that is summarised in figure 1.

Round 1 contained four open questions, designed using themes from the literature review aligned with the aims of the study, and each of the following two round of questions were then iteratively designed following the process of data analysis described below. The first round of questions firstly sought respondents' views on the resources available on the WHS for the development of a dark tourism product, and also whether they were aware of any current dark tourism offers on the site. These questions were posed to help the researchers to evaluate both respondent's knowledge of the site, but also to be able to place elements of the site on a continuum of dark tourism experiences (Miles, 2002; Stone, 2006). Questions were also asked about the respondents' views on the appropriateness of dark tourism as a new product offering, and whether they believed there was a market for this. These questions were designed to elicit responses relating to the ethical (Stone & Sharpley, 2008) issues associated with the development of dark tourism, and also to the commercialisation and development of tourism to WHS (Katahenggam, 2019; Nuryanti, 1996; Yi et al., 2018), particularly in regards to the authenticity of the tourism offer. Following the iterative design principles of Delphi studies (Avella, 2016; Donohoe & Needham, 2009; Sobaih et al., 2012) the two following rounds of questions were designed after reflecting on the answers from the first round and critically analysing them along with the literature on this topic, to elicit further responses and to examine areas of consensus or dissensus. For example, a question in the second round asked about what kinds of tourism activities (other than dark tourism) panellists through would be acceptable to develop locally. This helped to probe further into their views on tourism development to the WHS, given a strong negative response that emerged in the first round when asked about the development of dark tourism.

After each round of questions, a short report was sent to all participants outlining the findings of the analysis of that stage, and at the end, a more substantive report was sent outlining areas of

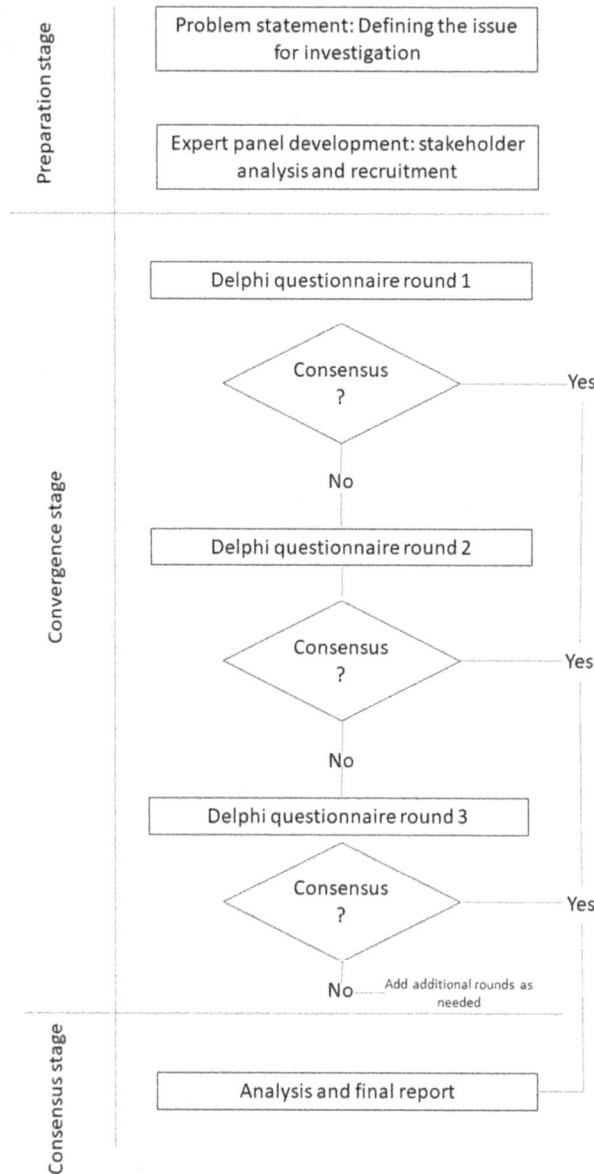

Figure 1. Delphi process (adapted from Donohoe and Needham (2009)).

consensus, to which participants were invited to send any additional points or to highlight any disagreements. At this final stage, no participants added additional information to the research. Each round of questions was sent out with two-week gaps, using Microsoft Forms for data collection and email for distribution of the survey links.

Panel membership is a key consideration for Delphi studies. This is an area in which the potential for researcher bias can be particularly high, as panellists must be chosen for their expert qualifications, ability to communicate on a topic, and willingness to participate in the study, which the researchers may not be best placed to evaluate, especially in advance (Avella, 2016; Sobaih et al., 2012). Sobaih et al. (2012) highlight that the criticisms of the Delphi method apply to all interpretive studies, despite this method's claims to provide some certainty, and that recognising and accounting for the subjectivity inherent in a Delphi design helps to add to its rigour. In the case of this study, the

Table 1. Participant information.

Participant number	Organisation	Position
1	National museum	Commercial Development Manager
2	Local history society	Principal
3	Open spaces management authority	Development Manager
4	University on the WHS	Commercial Director
5	Major commercial landlord	Marketing Manager
6	WHS managing authority	Development Director
7	University on the WHS	Tourism Researcher
8	Resident Association 1	Spokesperson
9	Tour Guide Business	Senior Manager
10	N/A	Maritime Historian
11	Destination Management Organisations	Senior Development Officer
12	Major Visitor Attraction within the WHS	Interpretation professional
13	Local Authority Heritage Organisation	Senior Manager
14	Transport provider	Manager
15	Resident association 2	Spokesperson

researchers have an excellent knowledge of the case, being members of the local community, as well as employees of an institution within the World Heritage Site itself.

Costa (2005) suggests that the accuracy of results from a Delphi panel increases as its size increases above 11 members, and that 15–20 may be an optimal number. Day and Bobeva (2005) report that most studies use between 15–35 people. In a study of stakeholder perspectives on interpretive methods for Canterbury Cathedral, part of a WHS in the United Kingdom, the panel was made up from ten respondents, and the findings were deemed to be rigorous given the expert nature of the respondents and their organisational attachments to the site. Using Mitchell *et al.'s* (1997) initial categorisation of stakeholders as *financial, moral, actual* and *potential*, a list of forty-seven possible participant organisations was created. These were further categorised, according to the same approach as *individuals, groups, neighbourhoods, organisations, institutions and societies.* Individuals were then identified who the authors believed would be best placed to give their opinions on the research, given their involvement in tourism or interpretation within the WHS. Fifteen participants were recruited for the first round of the study. Anonymised details of the respondents are given in Table 1.

With all Delphi studies, analytical techniques should be developed that meet the requirements of the study; there is no rule-book for data analysis. However, the nature of the method dictates that analysis should be iterative, as the 'waves' (Brady, 2015, p. 4) of data collection progressively inform the process of identifying consensus from the panel. To carry out this process, framework analysis (Ritchie & Spencer, 1994) was used as an appropriate analysis technique for relatively unstructured qualitative data. The framework approach involves four analytical steps: familiarisation with the data; identifying a thematic framework; applying the framework to the data; then interpreting the data afresh using this framework. This process is repeated until theoretical saturation occurs. This process leads to the identification of a final set of organising 'frames' for the analysis, which have been developed iteratively through immersion in, and analysis of, the qualitative data. In order to reduce bias in the analysis and increase the rigour of the findings, especially in terms of dependability (Walters, 2016), the process was followed independently by both authors, who then critically compared their analysis at each stage in order to deal with potential disagreements in interpretation. At the end of this framework analysis process, three frames of analysis were derived, which are used to structure the findings and discussion section, below.

Findings and discussion

Each of these frames of analysis derived from the methodology explained above is dealt with in turn in this section, where they are presented in a critical comparison to the literature reviewed for this research, in line with Hasson and Keeney's (2011) recommendations for increasing the rigour of the findings from qualitative Delphi panel studies. In the presentation of these findings, care has been taken not to identify the respondents in order to preserve the benefits of the anonymity offered by the Delphi approach, which can support respondents to engage with contentitious or controversial concepts, such as dark tourism (Avella, 2016; Habibi et al., 2014). The principle of anonymity is particularly important in a WHS site context, where multiple stakeholders are likely to be known to each other given the bounded nature of WHS designations.

F1: Perceptions of Dark Tourism's authenticity and suitability for the WHS

This first frame relates to the concerns expressed by respondents about the suitability of the development of dark tourism at the Greenwich WHS, because of the likelihood that it would not be authentic in nature (Katahenggam, 2019; Nuryanti, 1996; Yi et al., 2018). There was a general consensus from this Delphi panel that this was a major impediment to the development of dark tourism. One respondent explained that 'Promoting something less than truly authentic wouldn't stand up to much scrutiny … why do it somewhere like Greenwich with so much to offer that is both unique and authentic.' (P8). Another respondent said 'Not sure what there is that is authentic – it's not a battlefield or something like that' (P2). Respondents viewed the development of dark tourism as being a difficult interpretive challenge due to the lack of what they saw as authentic local dark heritage: 'You would need to develop some good story telling around dark tourism for WHS, I am not aware of any suitable stories or links to this for the site' (P13).

In line with Dann's (1994) and Stone and Sharpley's (2008) arguments about the sensitivities involved in developing dark tourism products, respondents expressed strong views about the suitability of dark tourism for the site. The WHS itself was perceived as being a serious place and not suitable for 'silly stories' (P2), with agreement that the 'lighter more entertainment orientated aspects of dark tourism would not have a place' (P7). Where respondents were more positive about future developments of this type, they were keen to explain that 'the most appropriate form of dark tourism for a WHS would be the 'dark side of existing history' type – i.e. authentic and place-based stories rather than the more generic ghost story/horror themes' (P7). Primarily though, despite some positive comments, there was a consensus that dark tourism would be 'Totally inappropriate' (P5) and 'certainly not something for families and children' (P13). It became apparent that the dark tourism products and experiences that the panel disliked were those at the lighter end of the dark tourism spectrum (Miles, 2002; Stone, 2006), and, more specifically, those with a more 'fright tourism' (Bristow, 2020) feel.

Respondents did not associate dark tourism with ideas of contested history or alternative interpretations of the site, which had been identified in previous research as being one way in which dark tourism can make a positive contribution (de-Miguel-Molina & Barrera-Gabaldón, 2019; Lemelin et al., 2013; Tunbridge & Ashworth, 1996). Panelists discussed the 'proud history' (P7) and 'rich history' of the site and the 'the main important attractions' (P11) that were the backbone of the visitor experience. Positive comments about dark tourism were most often qualified by saying that it should be kept separate and 'not infringe on the more mainstream offers or affect people's living and working environment' (P11). Specific suggestions were given for how this could be done such as: 'dark tourism experiences may be seen as more appropriately offered as a tailor made product for special groups' (P7). Research has previously explored tensions between the universal value attributed to WHS and contested local and other narratives for the heritage involved (Rakic & Chambers, 2008), but this panel presented a fairly orthodox interpretation of the Greenwich WHS, within which there is no obvious 'authentic' fit for dark tourism. This specific aspect of the panel's response is dealt with in more detail in F2.

F2: Specialist knowledge of the World Heritage Site

Panellists expressed very variable levels of knowledge about the heritage resources of the WHS and the different experiences that were available to tourists, as well as of the purpose of the site itself. This appeared to lead to differing attitudes to what was possible and desirable in terms of local tourism development. Despite the consensus in F1 on the dominant narratives and institutions of the site, there was little consensus evident on the more fine-grained detail of the site and its history, or of the purpose of the WHS itself.

When responding to questions about the possible dark tourism products or experiences on the WHS, some respondents were able to give detailed lists of specific sites and objects, and to draw links to phenomena such as the trade in slaves and sugar that characterised the British Empire during the time the site was built. These included executions and autopsies performed on site (P3); burial sites underneath buildings (P4) and in locations just outside the WHS boundaries (P7) and the bloodstained jacket of Admiral Nelson (P1), for example. However, most respondents discussed the WHS in more general terms as having a 'focus on maritime history' (P1) or being 'so rich in history and heritage' (P9). Many responses can be summed up by the statement: 'My history knowledge of the site is poor, but … ' (P4). Where respondents expressed a more detailed knowledge of the site, they tended to have a more positive perspective on the possible future development of dark tourism, but this was a minority view.

Community organisations and other stakeholders who are not located within institutions on the WHS itself were keen to highlight, throughout the rounds of questions, other, less-dominant local historical narratives. These included demands for 'more about the real, non-Royal Greenwich – less fluff' (P2), often with a focus on the industrial heritage of the area which includes telecommunications, shipping and shipbuilding that was linked more to the lives of residents and businesses.

The diversity of responses gathered together within this frame was successively probed through three rounds of questions, without a consensus developing in terms of content, but demonstrating the importance of understanding different stakeholders knowledge and valuation of the site. Dans and González (2019) argue that the successful management of tourism to heritage sites must include the balancing of competing views and values and Imon (2017) noted the challenges of doing this in an urban setting, such as this, where competing interpretations of cultural values co-exist. For future tourism development of any kind on the site to be viable, it is important that stakeholders share an understanding of the nature and heritage of the site. Although this research focused on dark tourism, any tourism developments which involved fresh interpretations of the site, or which focused on exploiting non-obvious locations or objects, would meet similarly diverse levels of support. Successful stakeholder management (Evans, 2002) of sites such as this is a pre-requisite for their effective management, and for the sustainable development of tourism (Liburd & Becken, 2017; Li et al., 2020; Rasoolimanesh & Jaafar, 2017; Timothy & Boyd, 2003).

F3: Specialist knowledge of local tourism

The final frame that was developed through the analysis of this panel's responses concerned respondents' specialist knowledge of the scope and scale of local tourism development. This emerged as a topic during the iterative process of questions, analysis and reflection outlined in the methods section, above. A number of panellists' responses in the first round of the research suggested that there were negative perceptions of tourism development *per se*, which were not expected given the fact that all of the respondents were stakeholders of a WHS that formed a significant international tourism destination. Through successive rounds of questions, it became clear that panellists' level of knowledge of the nature of local tourism tended to affect their views on whether *any* form of tourism development locally was desirable. Tourism is a core element of the WHS programme (Cassel & Pashkevich, 2014), and, with some significant exceptions, the panel expressed consensus that it was important locally, when this topic was investigated through successive rounds.

As you would expect from the stakeholders of an international tourism destination, panellists expressed some detailed knowledge about the current levels and qualities of local tourism. There was a consensus that the market is 'dominated by heritage visitors' (P4), which would be expected for a WHS, but that the offer of the site 'currently meets a limited demographic' (P6) who are very present on the site during the day, but that the destination is much quieter at night 'when tourists don't visit so much or residents go elsewhere for the evening' (P11). Mostly, the core museum and attractions of the site were mentioned by panellists, although there were occasional mentions of the other service offered to tourists such as the 'shops, markets and restaurants' (P9).

From a minority of respondents, the main issue that they were keen to put forward as a reason for not developing dark tourism on the site was not the nature of dark tourism itself, but a more general concern about the quality and volume of tourism to the WHS. Although no panellists used the term, these were concerns about overtourism (Dodds & Butler, 2019) that are common to many heritage tourism destinations (Adie et al., 2019; Seraphin et al., 2018). Stakeholders who did not represent institutions from within the site claimed that 'most of what tourists are told is populist' (P2) and that 'The tourist offer is at times already pretty debased and aimed at the lowest common denominator' (P8). This was clearly a critique of a perceived 'mass tourism model that cheapens the offer for tourists and has alienated residents' (P15). Much of this response was related to tourist numbers and the pressure this puts on local residents and services. One panellist stated that: 'There is a tipping point where the numbers begin to have a negative effect on the experience for all. At times Greenwich is already there' (P8). Another said that 'Public realm esp. planting is very poorly maintained, rubbish provision inadequate given the amount of street food. Provision of power points for food stalls in Cutty Sark Gardens also insufficient necessitating the need for diesel generators. This is all in the main entry point for tourists' (P15). One respondent posed the question: 'Bring in another area of tourism to what is already a very busy tourism site and destination?' before criticising the idea of developing dark tourism offers locally.

However, it was not the case that there was a consensus against developing any new form of tourism at all. As shown in F1, it was dark tourism that attracted particularly negative opinions, but the importance of tourism was very well recognised by the panel, and a number of suggestions were made about how this could look. Panellists with a close connection to the tourism industry expressed often quite well-thought-out suggestions for new forms of tourism to deal with the perceived biases and deficiencies of the current offer, as well as new demands from tourists:

> younger generations love to have access to many new experiences - immersive and interactive, exclusive (e.g. fine dining, sleepovers, etc.) that they can't access anywhere else but this is not something, which has been developed in Greenwich yet to a degree that it can offer quality and is available all your round. (P12)

> Here is opportunity to grow the offer for families with teenage children and young adults - especially in the evenings when there isn't much to do for younger people currently. There is also room to grow the offer for the more adventurous tourist - incl. physical experiences such as climbing, ice skating and other sports. (P11)

Other suggestions from panellists for future tourism development included wellness tourism, the MICE market, festivals and an enhanced food tourism offer.

Conclusions

This research has shown that, in the case of the Greenwich Maritime WHS, stakeholders do not favour the development of dark tourism. The analysis of these stakeholders' responses revealed three main issues that helped to explain this. Firstly, respondents tended to hold very negative views about dark tourism as a phenomenon. This meant that, when asked about the appropriateness of developing this type of tourism locally, or to identify potential resources that this offer could be built on, the most frequent response was that it should not take place at all, and that there would be very little of interest for dark tourists on the site, in any case. Probing this perspective, it became apparent that this was mostly influenced by panellists' perceptions of dark tourism as being dominated by fight tourism (Bristow, 2020) products such as ghost tours and escape rooms, and a sense that this lighter end of the dark tourism spectrum (Miles, 2002; Stone, 2006), was not in keeping with the more serious purpose of the site. This mirrors the findings of previous research (Dann, 1994; Stone & Sharpley, 2008) in which concerns have been raised about 'cashing in' on an expanding dark tourism market, and not dealing appropriately with the ethical issues this could raise. For the sustainable development of tourism at this WHS and in other similar locations, successful stakeholder engagement has been identified as a pre-requisite (Liburd & Becken, 2017; Li et al., 2020; Timothy & Boyd, 2003; Rasoolimanesh & Jaafar, 2017). When developing dark tourism, it is clear that stakeholder education and familiarisation would be necessary, to help stakeholders to understand the different shades of dark tourism, to make a more balanced judgement on their acceptability.

The second frame of analysis, however, saw a less consensual perspective emerge, relating to respondents' specialist knowledge of the WHS. When stakeholders had a more detailed knowledge of the heritage resources of the site, they were more able to consider linking these to potential dark tourism. Within this frame of analysis, the main issue affecting perceptions of dark tourism development at the WHS was the 'fit' of the developments with the identity of the site. This was not expressed using the 'universal' values of the WHS, as had been identified in previous research (Rakic & Chambers, 2008), instead it related to orthodox interpretations of the site as being concerned with royalty, the Navy and national prestige. These were the 'authentic' (Katahenggam, 2019; Nuryanti, 1996; Yi et al., 2018) values of the site, which dark tourism was seen as either undermining or contradicting. This frame showed that stakeholder perspectives on the potential development of dark tourism to complex, urban WHS with multiple heritage resources within it, are dependant on the level of knowledge that stakeholders have of these complex resources, and also on the dominant narratives of the site. This supports the views of Imon (2017), who argued that complex urban WHS present particular stakeholder issues and Dans and González (2019) who drew attention to the intricacies of balancing competing values in heritage tourism development.

The third frame of analysis was concerned with stakeholder attitudes towards tourism development at the WHS more generally. Two clear stakeholder perspectives emerged. Panellists who did not represent institutions within the WHS voiced concerns about potential overtourism (Dodds & Butler, 2019) impacts from developing additional tourism to the site. These included worries about congestion, pollution, litter and overcrowding, as well as negative resident attitudes. However, the majorities of panellists were very positive about local tourism growth, and were keen to suggest types of tourism that they saw as suitable. Although the impact of WHS status on tourism growth is mixed, mixed (Gao & Su, 2019; Mariani & Guizzardi, 2020), it is clear from this research that WHS stakeholders had a mostly positive attitude towards tourism and saw the growth of tourism locally as important for the future of the site. These more general attitudes towards tourism and its impacts were seen to have as important an influence over whether panellists were supportive of the idea of developing a new dark tourism offer, as perceptions of dark tourism itself.

Although effort was made to engage a wide range of stakeholders in this research, this qualitative research did not seek to gain universal coverage of stakeholders at the WHS, if this were even possible. Instead, the study sought to develop insights into the potential development of dark tourism which could be useful for other researchers in the fields of dark tourism and WHS tourism. Additionally, no national or international stakeholders were included in the research, to avoid large disparities of power or resources, but future research into dark tourism to WHS could include these powerful voices who can have a significant influence over tourism development. The WHS chosen for this research is in an urban setting, with multiple stakeholders from the visitor economy, but also from other sectors. Because of this, the findings of this research may have particular value for other WHS with complex stakeholder relationships.

The findings of this research will be valuable to heritage tourism professionals considering the possible future relationship between dark tourism and heritage. In particular, this research is placed into a WHS context, meaning that it has international implications for the future management of tourism to many sites associated with 'dark' pasts. For researchers, this study provides a consensus view from a significant group of heritage tourism stakeholders on the relationship between dark tourism and heritage.

Stakeholders of a WHS in a significant international tourism destination understand the importance and value of tourism to the site, and are keen to see this grow in the future. Despite this positive orientation, dark tourism was not viewed as an appropriate or attractive new offer. The reasons for this mostly arise from perceptions of it being an inauthentic, entertainment-based tourism offer, which conflicts with the serious and important purpose of the WHS. Although there was some dissent from this broad consensus point, it is clear that significant work would need to be done with stakeholders on the nature of dark tourism, and the resources upon which it could be developed, in order for this form of tourism development to be welcomed and sustainable.

Disclosure statement

No potential conflict of interest was reported by the author(s).

ORCID

James Kennell ⓘ http://orcid.org/0000-0002-7877-7843

References

Adie, B. A., & Amore, A. (2020). Transnational world heritage, (meta) governance and implications for tourism: An Italian case. *Annals of Tourism Research*, *80*, 102844.

Adie, B. A., Falk, M., & Savioli, M. (2019). Overtourism as a perceived threat to cultural heritage in Europe. *Current Issues in Tourism*. https://doi.org/10.1080/13683500.2019.1687661

Ashworth, G. (2004). Tourism and the heritage of atrocity: Managing the heritage of South Africa for entertainment. In T. V. Singh (Ed.), *New Horizons in tourism: Strange experiences and Stranger practices* (pp. 95–108). CAB International.

Ashworth, G. J., & Isaac, R. K. (2015). Have we illuminated the dark? Shifting perspectives on 'dark' tourism. *Tourism Recreation Research*, *40*(3), 316–325. https://doi.org/10.1080/02508281.2015.1075726

Avella, J. R. (2016). Delphi panels: Research design, procedures, advantages, and challenges. *International Journal of Doctoral Studies*, *11*(1), 305–321. https://doi.org/10.28945/3561

Biran, A., & Hyde, K. F. (2013). New perspectives on dark tourism. *International Journal of Culture, Tourism and Hospitality Research*, *7*(3), 191–198. https://doi.org/10.1108/IJCTHR-05-2013-0032

Biran, A., Poria, Y., & Oren, G. (2011). Sought experiences at (dark) heritage sites. *Annals of Tourism Research*, *38*(3), 820–841. https://doi.org/10.1016/j.annals.2010.12.001

Boyle, E. (2019, July). Borders of memory: Affirmation and contestation over Japan's heritage. *Japan Forum*, *31*(3), 293–312. Routledge. https://doi.org/10.1080/09555803.2018.1544582

Brady, S. R. (2015). Utilizing and adapting the Delphi method for use in qualitative research. *International Journal of Qualitative Methods*, *14*(5), 5. https://doi.org/10.1177/1609406915621381

Bristow, R. S. (2020). Communitas in fright tourism. *Tourism Geographies*, *22*(2), 319–337.

Bucior, C. (2020). History's priests, history's magicians: Exploring the contentious relationship between authorized heritage and ghost tourism in Gettysburg, Pennsylvania. *Journal of Heritage Tourism*, *15*(3), 328–340.

Cassel, S. H., & Pashkevich, A. (2014). World heritage and tourism innovation: Institutional frameworks and local adaptation. *European Planning Studies*, *22*(8), 1625–1640. https://doi.org/10.1080/09654313.2013.784605

Collins-Kreiner, N. (2016). Dark tourism as/is pilgrimage. *Current Issues in Tourism, 19*(12), 1185–1189. https://doi.org/10.1080/13683500.2015.1078299

Costa, C. A. (2005). The status and future of sport management: A Delphi study. *Journal of Sport Management, 19*(2), 117–142. https://doi.org/10.1123/jsm.19.2.117

Dale, C., & Robinson, N. (2011). Dark tourism. In Robinson, P., Heitmann, S., & Dieke, P. (Eds.), *Research themes for tourism* (Vol. 38, No.1, pp. 193–209). CABI.

Dann, G. (1994). Tourism the nostalgia industry of the future. In W. Theobald (Ed.), *Global tourism: The Next Decade* (pp. 55–67). Butterworth Heinemann.

Dans, E. P., & González, P. A. (2019). Sustainable tourism and social value at World heritage sites: Towards a conservation plan for Altamira, Spain. *Annals of Tourism Research, 74*, 68–80. https://doi.org/10.1016/j.annals.2018.10.011

Davey, M., & Gillespie, J. (2014). The great barrier reef world heritage marine protected area: Valuing local perspectives in environmental protection. *Australian Geographer, 45*(2), 131–145. https://doi.org/10.1080/00049182.2014.899025

Day, J., & Bobeva, M. (2005). A generic toolkit for the successful management of Delphi studies. *The Electronic Journal of Business Research Methodology, 3*(2), 103–116.

de Fauconberg, A., Berthon, P., & Berthon, J. P. (2018). Rethinking the marketing of World heritage sites: Giving the past a sustainable future. *Journal of Public Affairs, 18*(2), e1655. https://doi.org/10.1002/pa.1655

de-Miguel-Molina, M. and Barrera-Gabaldón, J. L. (2019). Controversial heritage: The Valley of the Fallen. *International Journal of Culture, Tourism and Hospitality Research, 13*(1), 128–143. https://doi.org/10.1108/IJCTHR-01-2019-0006

Dodds, R., & Butler, R. (2019). The phenomena of overtourism: A review. *International Journal of Tourism Cities, 5*(4), 519–528. https://doi.org/10.1108/IJTC-06-2019-0090

Donohoe, H. M., & Needham, R. D. (2009). Moving best practice forward: Delphi characteristics, advantages, potential problems, and solutions. *International Journal of Tourism Research, 11*(5), 415–437. https://doi.org/10.1002/jtr.709

Evans, G. (2002). Living in a World heritage City: Stakeholders in the dialectic of the universal and particular. *International Journal of Heritage Studies, 8*(2), 117–135. https://doi.org/10.1080/13527250220143913

Farkic, J. (2020). Consuming dystopic places: What answers are we looking for? *Tourism Management Perspectives, 33*, 100633. https://doi.org/10.1016/j.tmp.2019.100633

Fryer, P. (Producer). (2018). *The dark tourist. [TV series]*. Netflix.

Gao, Y., & Su, W. (2019). Is the World heritage just a title for tourism? *Annals of Tourism Research, 78*, 102748. https://doi.org/10.1016/j.annals.2019.102748

Habibi, A., Sarafrazi, A., & Izadyar, S. (2014). Delphi technique theoretical framework in qualitative research. *The International Journal of Engineering and Science, 3*(4), 8–13.

Hartmann, R. (2014). Dark tourism, thanatourism, and dissonance in heritage tourism management: New directions in contemporary tourism research. *Journal of Heritage Tourism, 9*(2), 166–182. https://doi.org/10.1080/1743873X.2013.807266

Hasson, F., & Keeney, S. (2011). Enhancing rigour in the Delphi technique research. *Technological Forecasting and Social Change, 78*(9), 1695–1704. https://doi.org/10.1016/j.techfore.2011.04.005

Hyde, K. F., & Harman, S. (2011). Motives for a secular pilgrimage to the Gallipoli battlefields. *Tourism Management, 32*(6), 1343–1351. https://doi.org/10.1016/j.tourman.2011.01.008

Imon, S. S. (2017). Cultural heritage management under tourism pressure. *Worldwide Hospitality and Tourism Themes, 9*(3), 335–348. https://doi.org/10.1108/WHATT-02-2017-0007

Isaac, R. K., & Çakmak, E. (2014). Understanding visitor's motivation at sites of death and disaster: The case of former transit camp Westerbork, the Netherlands. *Current Issues in Tourism, 17*(2), 164–179. https://doi.org/10.1080/13683500.2013.776021

Ivanova, P., & Light, D. (2018). 'It's not that we like death or anything': Exploring the motivations and experiences of visitors to a lighter dark tourism attraction. *Journal of Heritage Tourism, 13*(4), 356–369. https://doi.org/10.1080/1743873X.2017.1371181

Kamber, M., Karafotias, T., & Tsitoura, T. (2016). Dark heritage tourism and the Sarajevo siege. *Journal of Tourism and Cultural Change, 14*(3), 255–269. https://doi.org/10.1080/14766825.2016.1169346

Katahenggam, N. (2019). Tourist perceptions and preferences of authenticity in heritage tourism: Visual comparative study of George Town and Singapore. *Journal of Tourism and Cultural Change, 1–15.* https://doi.org/10.1080/14766825.2019.1659282

Kennell, J., Šuligoj, M., & Lesjak, M. (2018). Dark events: Commemoration and collective memory in the former Yugoslavia. *Event Management, 22*(6), 945–963.

Kim, H., Oh, C. O., Lee, S., & Lee, S. (2018). Assessing the economic values of World heritage sites and the effects of perceived authenticity on their values. *International Journal of Tourism Research, 20*(1), 126–136. https://doi.org/10.1002/jtr.2169

Kwiatek-Sołtys, A., & Bajgier-Kowalska, M. (2019). The role of cultural heritage sites in the creation of tourism potential of small towns in Poland. *European Spatial Research and Policy, 26*(2), 237–255. https://doi.org/10.18778/1231-1952.26.2.11

Landorf, C. (2009). Managing for sustainable tourism: A review of six cultural World heritage sites. *Journal of Sustainable Tourism*, 17(1), 53–70. https://doi.org/10.1080/09669580802159719

Leask, A., & Fyall, A. (2001). World heritage site designation: Future implications from a UK perspective. *Tourism Recreation Research*, 26(1), 55–63. https://doi.org/10.1080/02508281.2001.11081177

Lemelin, H., Whyte, R. K., Johansen, K., Higgins Desbiolles, F., Wilson, C., & Hemming, S. (2013). Conflicts, battlefields, indigenous peoples and tourism: Addressing dissonant heritage. In: Warfare tourism in Australia and North America in the twenty-first century. *International Journal of Culture, Tourism and Hospitality Research*, 7(3), 257–227. https://doi.org/10.1108/IJCTHR-05-2012-0038

Lennon, J., & Foley, M. (2000). *Dark tourism: The attraction of death and disaster*. Thomson.

Yiping Li, Chammy Lau, & Ping Su. (2020). Heritage tourism stakeholder conflict: A case of a World Heritage Site in China. *Journal of Tourism and Cultural Change*, 18(3), 267–287. https://doi.org/10.1080/14766825.2020.1722141

Liburd, J. J., & Becken, S. (2017). Values in nature conservation, tourism and UNESCO World Heritage Site stewardship. *Journal of Sustainable Tourism*, 25(12), 1719–1735.

Light, D. (2017). Progress in dark tourism and thanatourism research: An uneasy relationship with heritage tourism. *Tourism Management*, 61, 275–301. https://doi.org/10.1016/j.tourman.2017.01.011

Lin, V. S., & Song, H. (2015). A review of Delphi forecasting research in tourism. *Current Issues in Tourism*, 18(12), 1099–1131. https://doi.org/10.1080/13683500.2014.967187

Magee, R. (2018) *Servicescape management at heritage tourism sites: From dark tourism sites to socially-symbolic servicescapes*. [Doctoral Thesis, University of Ulster]. https://pure.ulster.ac.uk/files/12687265/2018MageeRCPhD.pdf

Mariani, M. M., & Guizzardi, A. (2020). Does designation as a UNESCO world heritage site influence tourist evaluation of a local destination? *Journal of Travel Research*, 59(1), 22–36. https://doi.org/10.1177/0047287518821737

Maritime Greenwich World Heritage Site. (2019). *Home page*. http://www.greenwichworldheritage.org/

Miles, S. (2002). Auschwitz: Museum interpretation and Darker tourism. *Annals of Tourism Research*, 29(4), 1175–1178. https://doi.org/10.1016/S0160-7383(02)00054-3

Mitchell, R. K., Agle, B. R., & Wood, D. J. (1997). Toward a theory of stakeholder identification and salience: Defining the principle of who and what really counts. *Academy of Management Review*, 22(4), 853–886. https://doi.org/10.5465/amr.1997.9711022105

Murtagh, B., Boland, P., & Shirlow, P. (2017). Contested heritages and cultural tourism. *International Journal of Heritage Studies*, 23(6), 506–520. https://doi.org/10.1080/13527258.2017.1287118

Nuryanti, W. (1996). Heritage and postmodern tourism. *Annals of Tourism Research*, 23(2), 249–260. https://doi.org/10.1016/0160-7383(95)00062-3

Podoshen, J. S., Venkatesh, V., Wallin, J., Andrzejewski, S. A., & Jin, Z. (2015). Dystopian dark tourism: An exploratory examination. *Tourism Management*, 51, 316–328. https://doi.org/10.1016/j.tourman.2015.05.002

Powell, R., & Kennell, J. (2016). Dark cities? Developing a methodology for researching dark tourism in European cities. In *Tourism and culture in the age of innovation* (pp. 303–319). Springer, Cham.

Raine, R. (2013). "A dark tourist spectrum. *International Journal of Culture, Tourism and Hospitality Research*, 7(3), 242–256. https://doi.org/10.1108/IJCTHR-05-2012-0037

Rakic, T., & Chambers, D. (2008). World heritage: Exploring the tension between the national and the 'universal'. *Journal of Heritage Tourism*, 2(3), 145–155. https://doi.org/10.2167/jht056.0

Rasoolimanesh, S. M., & Jaafar, M. (2017). Sustainable tourism development and residents' perceptions in World heritage site destinations. *Asia Pacific Journal of Tourism Research*, 22(1), 34–48. https://doi.org/10.1080/10941665.2016.1175491

Richards, G. (2001). *Cultural attractions in Europe*. CABI publishing.

Ritchie, J., & Spencer, L. (1994). Qualitative data analysis for applied policy research by Jane Ritchie and Liz Spencer. In A. Bryman & R. G. Burgess (Eds.), *Analysing qualitative data'* (pp. 173–194). Routledge.

Roberts, C. (2018). Educating the (dark) Masses: Dark tourism and Sensemaking. In Philip R. Stone, Rudi Hartmann, Tony Seaton, Richard Sharpley, & Leanne White (Eds.), *The Palgrave Handbook of dark tourism studies* (pp. 603–637). Palgrave Macmillan.

Seraphin, H., Sheeran, P., & Pilato, M. (2018). Over-tourism and the fall of Venice as a destination. *Journal of Destination Marketing & Management*, 9, 374–376. https://doi.org/10.1016/j.jdmm.2018.01.011

Sharpley, R. (2005). Travels to the edge of darkness: Towards a typology of dark tourism. In C. Ryan (Ed.), *Taking tourism to the limits: Issues, concepts and managerial perspectives* (pp. 217–228). Elsevier.

Sobaih, A. E. E., Ritchie, C., & Jones, E. (2012). Consulting the oracle?: Applications of modified Delphi technique to qualitative research in the hospitality industry. *International Journal of Contemporary Hospitality Management*, 24(6), 886–906.

Stone, P. (2006). A dark tourism spectrum: Towards a typology of death and macabre related tourist sites, attractions and exhibitions. *Tourism*, 54(2), 145–160.

Stone, P. (2013). Dark tourism scholarship: A critical review. *International Journal of Culture, Tourism and Hospitality Research*, 7(3), 307–318. https://doi.org/10.1108/IJCTHR-06-2013-0039

Stone, P., & Sharpley, R. (2008). Consuming dark tourism: A thanatological perspective. *Annals of Tourism Research*, 25, 202–227. https://doi.org/10.1016/j.annals.2008.02.003

Su, M. M., & Wall, G. (2012). Global–local relationships and governance issues at the Great Wall World Heritage Site, China. *Journal of Sustainable Tourism*, *20*(8), 1067–1086.

Su, M. M., Wall, G., Wang, Y., & Jin, M. (2017). Multi-agency management of a World heritage site: Wulingyuan Scenic and historic interest area, China. *Current Issues in Tourism*, *20*(12), 1290–1309. https://doi.org/10.1080/13683500.2016.1261810

Tarlow, P. (2005). Dark tourism: The appealing 'dark' side of tourism and more. In M. Novelli (Ed.), *Niche tourism: Contemporary issues, trends and cases* (pp. 47–57). Routledge.

Timothy, D. J., & Boyd, S. W. (2003). *Heritage tourism*. Pearson Education Ltd.

Tunbridge, J. E., & Ashworth, G. J. (1996). *Dissonant heritage: The management of the past as a resource in conflict.* John Wiley & Sons.

UNESCO. (2020). *World Heritage List.* https://whc.unesco.org/en/list/

Visit Greenwich (2018). Annual report 2017-18. Visit Greenwich

Walters, T. (2016). Using thematic analysis in tourism research. *Tourism Analysis*, *21*(1), 107–116. https://doi.org/10.3727/108354216X14537459509017

Yi, X., Fu, X., Yu, L., & Jiang, L. (2018). Authenticity and loyalty at heritage sites: The moderation effect of postmodern authenticity. *Tourism Management*, *67*, 411–424. https://doi.org/10.1016/j.tourman.2018.01.013

Touring female crime: power and perceptions

Bailey Ashton Adie ⓘ and Esther J. Snell

ABSTRACT

Popular interest in crime is substantial and longstanding, driving the development of crime-based dark tourism attractions. The appeal of these sites can partly be explained through the understanding of functions of transgression as tours provide their audiences with infotainment. These representations of crime both reflect and shape social and cultural perceptions of the nature of offending and victimization. There is, however, a significant gap in relation to the discussion of these crime-based dark tourism activities with almost no engagement with gender at these sites. To fill this gap, this paper presents a conceptual discussion on tourism to sites of female criminal activity, drawing parallels to similar male crime locations. Examination of online advertising for murder walking tours in the UK reveals gendered power dynamics wherein traditional, western gender roles are enforced through the removal of agency from women who engage in more violent crimes while simultaneously fetishizing women as victims of violence, especially sexual. This is evident in the absence of female serial killers within organized dark tours, which often focus specifically on this sexual violence. Thus, the tourist activities that revolve around dark heritage sites, especially those that deal with violent criminal activity, reinforce gendered stereotypes around 'acceptable' transgression.

Introduction

While interest in murders and murderers has a long-standing history, in recent years, the commodification and consumption of narratives and depictions of murder have greatly expanded in diversity, quantity, and popularity (Haggerty & Ellerbrok, 2011; Oleson, 2006). In recent years, the ways in which murder can be consumed have expanded to include ever-more immersive and interactive pursuits. One of these consumption activities, the murder tour, allows the consumer to engage in, learn about, and vicariously observe and participate in acts of murder. Murder tours require the consumer to actively participate in the recounting of the case: to attend and view the murder sites, imagining themselves following in the footsteps of the victim, offender and investigator, pouring over evidence, and trying to solve the case. In this way, murder has been reformed as fun and entertaining, at least for those safely looking in from the outside, years later. It is noticeable, however, that these murder tours largely focus on cases in which men are the murderers and females their victims. As in society more widely, the agency of women as killers, and serial killers, in these tours is largely overlooked in favour of their male counterparts. While female perpetrators have a noted absence within actual dark tourism activities, there is an even larger gap in the dark tourism literature.

Dark tourism as an area of study has existed for a little over 20 years, but the number of studies which specifically discuss murder-based dark tourism activities are limited (Dalton, 2015; Gibson, 2006; Heidelberg, 2015; Huey, 2011; Kim & Butler, 2015; Lennon, 2010; Powell & Iankova, 2016; White, 2013; Wilbert & Hansen, 2009). In fact, Gibson (2006), in his call for future research into the topic, emphasized the need for a better understanding of serial killer tourism. Furthermore, none of the previous works fully engage with the gender dynamic of the murder narrative provided, although Huey (2011) and Wilbert and Hansen (2009) do touch on it. Therefore, in order to address these significant gaps in the literature, this article presents a discussion of the presentation of female offending in murder tours. In order to illustrate these gendered narratives, a selection of tours was gathered through the use of an online keyword search which specifically focussed on murder walking tours. Additionally, to further narrow down the results, only verifiable historical murders were included, and the tours were required to present murder as their dominant theme. This resulted in the identification of 51 murder walking tours, 49 of which were regularly offered and two that were singular events. The websites for these tours were then analysed, taking into consideration both murderers and victims highlighted as part of the tour. In addition, the online presence of sixty-five murder cases (representing sixty-nine known murderers: 25 male and 44 female) were identified to show the potential pool for murder tour content.

Through this analysis, it becomes apparent that the focus of these tours is almost exclusively male. Furthermore, while the victims of these male murderers are not always female, overall the cases represented in the murder tours represent the predisposition of modern society to cast men and women in the proactive and passive roles respectively. While it is not suggested that this is done deliberately, nevertheless, in terms of interest, marketing, and understandings of both the nature and frequency of murder, these murder tours reflect modern preconceptions of stereotypical gender roles, unknowingly reinforcing gender stereotypes.

Murder, the media, and gender: a (brief) history

The public's interest in murder is nothing new. Considered one of the most serious offences, murder was officially punished by execution in England until 1969. In the eighteenth-century, public execution of murderers drew not only large crowds but also significant numbers of people willing to pay for narratives of the heinous acts. Pamphlets, newspaper articles, broadsides, and court accounts were all popular during the period. These narratives reveal considerable interest in female killers. The crimes of murderers Mary Blandy, Elizabeth Jeffryes, Mary Edmondson, and Elizabeth Brownrigg were some of the biggest cases of the period. Part of the horror of these crimes appears to lie in the intimate nature of the relationship between the murderer and victim, and the nexus of threat within the home:

> 'What a Shudder must human Nature receive, when it recollects there is no Place where security may be depended upon, but at the same Time Persons are barring their Doors from Thieves without, they are inclosing worse Enemies within' (Anon, 1752).

The Victorians also loved tales of murder, a fact recognized by the newspaper proprietors of the day. By this time, however, the attitude towards female killers had changed somewhat from regarding them as unnatural monsters to, more often, exploited women deserving of some understanding. Consequently, female murderers were often shown mercy and sentenced to prison rather than execution. The case of Florence Maybrick, convicted in 1889 for murdering her husband, is an example of this.

As Fleming (2007, p. 277) observes, 'few crimes elicit as significant a public response as serial murder.' Some cases are so widely publicized and featured in media forms, they become 'mega-cases' (Fleming, 2007; Soothill et al., 2002). However, in modern times, this representation appears to be largely synonymous with men. Soothill et al.'s (2002) study of the UK newspaper coverage of mega-cases from 1977 to 1997 reveals only two which featured female killers: The Moors murders committed by Myra Hindley and Ian Brady, and those perpetrated by Rosemary and Fred West. None of the mega-cases featured female killers acting alone. For example, the crimes of Beverley Allitt, who was convicted in 1993 for the murder of 4 children and the grievous bodily harm of 6 more, did not attain mega-case status. Soothill et al. (2002) argue that not all homicides are treated in the same way and that there are hierarchies of cases. The strongest predictor of whether a crime will receive press attention are the number of victims, followed by the number of offenders and whether the offender utilized an unusual method of killing. The murder of women and children are also more likely to be reported, with females featuring significantly as victims, not protagonists (Soothill et al., 2002). The victims identified in the mega-cases included Julie Dart, Rachel Nickell, Lin and Megan Russell, Jasmine Beckford, and the multiple female victims of the 'Railway murderers' John Duffy and David Mulcahy.

The cultural representation of these offences is important. Not only do they serve social functions (Fleming, 2007) but they become the medium through which murder is both viewed and presented in the media (Soothill et al., 2002). Over time such prominent cases become 'point[s] of reference for commentary and discussion in response to later murders' (Soothill et al., 2002, p. 420). In short, as narratives of murder are created and consumed, those tropes inform and set boundaries for what is considered of interest and how subsequent cases are understood and interpreted. Media representations of what constitutes both a murder, and one that is newsworthy, serve to inform and impact on what cases are deemed noteworthy and, more broadly, their analysis and future representations. As Haggerty and Ellerbrok (2011, p. 6) state, 'serial killing is intimately tied to its broader social and historical settings'. The same is true for its cultural representation. Discourse reflects social, political, and cultural resonances, revealing the themes that are of interest and concern to the audiences as well as what the producers think audiences will want to consume. Those imperatives evolve over time. If female killers are not featured as prominently as men in modern murder tours, then the reason must partly lie in social understandings of gender and crime and its use and representation in the modern world. However, a distorted representation could lead to mistaken understandings of homicide as a social problem, leading to the proliferation of inaccurate stereotypes (Gruenewald et al., 2009).

Dark tourism

While travel to sites associated with death and tragedy is not a new trend (Gibson, 2006; Lennon & Foley, 2000; Seaton, 1996; Stone, 2012; Stone & Sharpley, 2008), the concept of dark tourism is fairly recent, with its roots in heritage tourism (Light, 2017). The term 'dark tourism' was first used by Foley and Lennon (1996, p. 198) who defined it as 'the presentation and consumption (by visitors) of real and commodified death and disaster sites.' Situated in the same special issue on the general topic of 'dark tourism' was Seaton's (1996, p. 240) conceptualization of thanatourism, which 'is travel to a location wholly, or partially, motivated by the desire for actual or symbolic encounters with death, particularly, but not exclusively, violent death.' Thanatourism, as can be noted, includes death as a motivating factor, which is not inherent in Lennon and Foley's dark tourism concept. However, Lennon and Foley (2000), when expanding on the concept, stressed that dark tourism is a post-modern construct which was a direct result of the 'anxiety' towards the modern world which these sites evoke (Foley & Lennon, 1996). Light (2017) notes that, when originally conceived, the main difference between these two concepts was that dark tourism was inherently focused on the supply side of dark sites while thanatourism was based on the demand-side aspects of tourist motivation and behaviour.

Stone (2013, p. 307) further refined the concept of dark tourism as 'tourist encounters with spaces of death or calamity that have perturbed the public consciousness, whereby actual and recreated places of the deceased, horror, atrocity, or depravity, are consumed through visitor experiences.' In an effort to expand on the complexity inherent in the dark tourism concept, Stone (2006) developed a dark tourism spectrum, ranging from the darkest to lightest of sites, which identified seven specific sub-types of dark sites: dark fun factories (i.e. the London Dungeon), dark exhibitions, dark dungeons (i.e. Alcatraz in San Francisco, USA), dark resting places (i.e. Lafayette Cemetery No. 1 in New Orleans, USA), dark shrines (i.e. Elvis's grave in Graceland, USA), dark conflict sites, and dark camps of genocide. While dark tourism has advanced as a concept since 1996, there is still some criticism that has arisen in response to it. Some is due to the concept's focus on Western contexts (Yoshida et al., 2016) or the use of the word 'dark' in general as Hartmann (2014) has noted that it may be difficult if not impossible to translate the meaning of the word in other languages. Seaton (2009) has also questioned the use of the word 'dark' albeit for a different reason. Seaton (2009, p. 525) has referred to the use of the word 'dark' as equating to a moral judgement wherein sites of this type are inherently transgressive, which in turn positions non-dark sites as 'light' and morally acceptable. Light (2017) has argued that dark tourism may not even be a distinct subtype of tourism from a demand perspective as dark tourists are often indistinguishable from heritage tourists.

Regardless of the criticism, there has been considerable research on the topic, covering sites of genocide (Ashworth & Tunbridge, 2017; Friedrich et al., 2018), mass murder (Dalton, 2015; Kang et al., 2012), disasters (Yoshida et al., 2016), terrorism (Frew, 2017), slavery (Hanna et al., 2018; Rice, 2009), battles and associated military conflict (Baldwin & Sharpley, 2009; Vanneste & Winter, 2018), assassinations (Foley & Lennon, 1996), as well as lighter dark tourism destinations like the London or York Dungeon (Ivanova & Light, 2018; Stone, 2009). However, there has been little engagement with crime as a dark tourism theme. Crime and dark tourism, when discussed together, often focus on 'crimes against humanity' as opposed to crimes against individuals, as was visible in Dalton's (2015) work, *Dark Tourism and Crime*. His case studies included sites of genocide, war crimes, atrocities carried out by political regimes, and terrorism. Only two case studies specifically engage with criminal activity: the memorial to John Lennon in Central Park, New York, and the Port Arthur memorial in Australia. In both instances, the sites memorialize the victims of these crimes, and, in the case of Port Arthur, not only is visitation not promoted as part of the site tour but questions from visitors about the shooting are actively discouraged. In his conclusions, Dalton (2015) does mention another crime-related dark tourism site, Snowtown in Australia, where several murder victims were found. However, this is treated as an 'unwanted' site as, although there is noted benefit to the local town, there is no political will to create an official tourism destination.

Touring the scene of the crime

Snowtown is an interesting case as, regardless of infrastructure, there is a consistent influx of dark tourists interested in the murders (Kim & Butler, 2015). This illustrates what Timothy (2018, p. 386), has noted in his study on dark heritage in the USA, namely that 'people are morbidly fascinated by mass murders and serial killings.' However, violent crime, and more specifically murder, are discussed much less frequently than other forms of dark tourism within the literature. Dark tourism at sites of murder often occurs following media coverage of the event. This is not a new trend, as was visible in the ad-hoc visitation which occurred at several murder locations in the US and UK in the nineteenth century and continues to this day (Gibson, 2006). However, as these sites develop extemporaneously, they can often cause issues for local governments whose constituents may be personally connected to the victims or have other memories of the criminal events (Heidelberg, 2015; Kim & Butler, 2015; Lennon, 2010). Lennon (2010) highlights that management of recent sites often includes demolition to deter visitation, emphasizing the importance of temporal distance from the event when considering development of murder-related dark tourism sites. Additionally, while not all murder sites become lasting dark tourism destinations, those that do may face issues when local governments refused to manage the visitor narrative, if only to mitigate potential damage to the local reputation. This is visible in Amityville in New York, whose public image has been eclipsed by the Amityville Horror narrative, in part due to its popularity within popular culture (Heidelberg, 2015). Local control of the narrative can also avoid the development of dark tourism *kitsch* which may be viewed as insensitive responses to criminal events (Kim & Butler, 2015).

One way to shift the interest from a geographical location is through the movement of artefacts to museum contexts. Murder within a museum provides a blend of education and entertainment, creating more controlled narratives which allow visitors 'to explore fears in ways that are not only physically and ontologically safe but also allow for the conversion of fear in pleasure' (Huey, 2011, p. 397). While there are multiple crime museums around the world, the most common murder-based, organized dark tourism activity which is discussed in the literature are walking tours (Powell & Iankova, 2016; Wilbert & Hansen, 2009). While both Powell and Iankova (2016) as well as Wilbert and Hansen (2009) discuss Jack the Ripper tours in London, Powell and Iankova (2016) present these as part of the dark tourism fabric of London, which also includes attractions like the London Dungeon, Madame Tussaud's, and the Tower of London. They note that, in London at least, dark tourism is increasing in popularity and 'much more mainstream and commercialized than in many other comparable cities' (Powell & Iankova, 2016, p. 349). However, this commercialization would appear to be problematic as it has, in its memorialization of the unknown perpetrator, erased the identities of the victims (Wilbert & Hansen, 2009). While Jack the Ripper is infamous both for his actual crimes as well as his fictional characterizations, other murders or serial killings may be less well embedded in the global, public consciousness but still of interest to certain groups who are curious about this niche area of dark murder tourism.

The popularity of dark tourism activities around murder, especially Jack the Ripper, was visible in the 51 tours analysed of which 29 were located in London and 23 featured the crimes of Jack the Ripper, making this the most commonly covered case and emphasizing the enduring popularity of the Whitechapel murders. While the original 'Jack the Ripper Tour' has been running since 1982 (Discovery Tours & Events, 2020a), the variety of tours has expanded significantly, with some providing food and drink while another blends the real-life murders with the exploits of the fictional Sherlock Homes. One tour comes with 'Ripper Vision', where 5-foot projections of stills and moving film enhance the experience (TripAdvisor, 2020). This fascination with serial murder is also visible in the USA where dark tourists can visit sites associated with the Zodiac Killer, Jeffrey Dahmer, H.H. Holmes, and the Manson Family. For a more global experience, there is even a serial killer app which allow users to find serial murder sites around the globe and potentially develop self-directed murder tours (Bolan & Simone-Charteris, 2018). These types of attractions often sell themselves as family fun that is entertaining, educational, and well-researched. Using Ferrel's (2004) conceptualization of crime as a response to boredom with the modern world, murder tours allow individuals to consume mediated vicarious engagement in criminal offending that fits the context of pre-arranged, exciting yet safe encounters with criminal actions and their perpetrators, as was observed in Huey (2011). This pursuit of excitement is, Ferrell (2004) argues, a desire for human engagement. Following the footsteps of societies' most despised and feared, but also fascinating, killers allows for a voyeuristic participation that is titillating but ultimately safe.

Whither the wicked women?

While the popularity of cultural representations of murder might be explainable, why male killers, especially serial killers, are covered more frequently than female killers is not so evident. It could, perhaps be linked to the academic pre-occupation with male killers, which has led to traditional typologies of killers that were, due to their restricted definitions, neglectful of women. In fact, the very definition of serial murder, which emphasized sexual motivation (an aspect women do not tend to share), served to exclude female killers (Farrell et al., 2011). The result of this is that female actions have 'been misconstrued, overlooked, and underestimated' (Harrison et al., 2015, p. 384). It is unlikely, however, that it is just this traditional lack of understanding of female killers that has led to their neglect in modern murder tours. A growing body of research focusing on female killers has led both to the development of broader definitions and typologies that better include women, as well as greater knowledge of their motivations, *modi operandi*, and targets. Studies suggest that while men tend to target strangers, women often choose victims who are older than those selected by men, vulnerable, and for whom they have a care-giving role. Women take longer to entrap their victims; are active for longer periods of time; accrue more fatalities; are less ostentatious, sadistic or sexual in the act, which is frequently achieved via poison; and their motivation is often financial. They also tend to be educated, married at some point, and Caucasian (Farrell et al., 2011; Farrell et al., 2013; Harrison et al., 2015; Hellen et al., 2015).

Similarly, while it has been often noted that female killers are rarer than their male counterparts (Farrell et al., 2011), there still exist a significant number of them who could populate murder tours. Between 1674 and 1913, the Central Criminal Court in London (the Old Bailey) heard 5382 cases of homicide, of which 1192 defendants were women. Of those trials, 494 women were found guilty of murder. From a low of nine trials in the 1700s, the 1840s and 1850s witnessed a noticeable increase in the numbers of female defendants, with 62 and 82 trials respectively, with numbers reaching a high of 126 in the 1880s before falling to 27 in the 1910s (Hitchcock et al., 2012). There are numerous other notable cases tried around the UK that could be used to populate several murder tours. To illustrate, Mary Ann Cotton, executed in Durham in 1873, was thought to have used arsenic-laced tea to poison up to 21 people, including three of her husbands and 11 of her children. Curious sightseers still visit her house in West Auckland, and the Beamish museum claims to have her teapot, enticing potential visitors (Beamish, 2016; Hutchinson, 2016). Still, there is no evidence that she is included in any organized tour. In another example, Kate Webster, an Irish servant, murdered her mistress in 1879; proceeded to dismember, burn and scatter her remains; and then assumed the mistress's identity for two weeks. This cause-celeb resulted in pamphlets, newspaper reports, and ballads, and yet, despite the rediscovery of the victim's head in 2010, today is largely forgotten.

Notwithstanding the numerous potential cases, most of the killings featured in the online marketing of UK murder tours are those committed by men. Of the 51 tours analysed, only five commercial and regularly organized tours, and a further one unique event, advertised murders committed by women in their online marketing (See Table 1). Together these tours featured six female murderers, five of whom were directly named: Myra Hindley, Ruth Ellis, baby farmers Amelia Sach and Annie Walters, and Constance Kent (constituting 11% of the 44 female murderers identified in the online sample). Two tours also featured the murders committed by Christiana Edmunds in 1871. Edmunds, the so-called Chocolate Box Poisoner or Chocolate Cream Killer, poisoned numerous people, including four-year-old Sidney Baker, with strychnine-laced chocolates. However, neither of the Brighton-based tours directly named Edmunds or identified her gender in their marketing. Although this was a significant case at the time and one which featured in the 1970 television series, *Wicked Women*, today it is largely forgotten, and we can see that these tours preferred to utilize Edmund's criminal moniker. Of course, to have a pseudonym at all is to suggest the original public appeal of the case. Two further female murderers Mary Pearcey (proposed as a possible Jack the Ripper suspect) and Mary Ann Britland were both discussed on the supplementary webpages belonging to two tours. However, neither women were identified as being a part of the actual walking tours or mentioned in their direct marketing on the homepage.

These examples stand in contrast to the many murder tours in which the names of featured male killers are listed prominently. Of twenty-five UK male murderers analysed twelve (48% of the online sample of murderers) were directly named in murder tour marketing with two more advertised by their criminal monikers. Such marketing utilizes the names associated with prominent and renowned cases to draw the attention of potential customers. Therefore, it is noteworthy that these tours predominantly promote male-dominated cases to the exclusion of female ones. It indicates that female-focused cases are not believed to be as prominent in the popular consciousness or as capable of attracting customers as male murderers. In many instances, as is the case with Jack the Ripper in the UK and Jeffrey Dahmer and Ted Bundy in the USA, the tours may solely be about the murderous exploits of one male killer. In comparison, no female murderer in the sample was given

Table 1. UK Murder Walking Tour Sample.

Tour Location	Historical Murderers Featured	Tour Name	Female Murderers Featured: # and Name (N) or Pseudonym (P)	Female Victims Featured: Named (N) or Anonymous (A)
London	Whitechapel Murders only	**Jack the Ripper the Original Terror Tour**	1 (N)	Yes (N)
		Jack the Ripper Guided Walks in London	0	Yes (A)
		Jack the Ripper interactive Ghost Hunts		Yes (A)
		Jack the Ripper and Sherlock Holmes Tour of Haunted London	0	No
		Jack the Ripper and Haunted London Tour	0	No
		Jack the Ripper Tour, with Ripper Vision (comparison with Yorkshire Ripper)	0	No
		Free Tours by Feet	0	No
		Jack the Ripper Tour: On the trail of a killer	0	No
		Free Jack the Ripper Tours	0	Yes (N)
		Jack the Ripper Murder Mystery Tour	0	No
		Art and Murder in Whitechapel	0	No
		Jack the Ripper, Haunted London and Sherlock Holmes Tour	0	Yes (N)
		Jack the Ripper's Wicked London	0	Yes (A)
		Murder, Medicine and Missionaries: The hidden East End	0	No
		Jack the Ripper and Haunted London Tour (by Evan Evans)	0	No
		Jack the Ripper Tour	0	No
		Jack the Ripper Walking Tour	0	No
		London: Grim Reaper Walking Tour	0	Yes (A)
		The Jack the Ripper Tour: A walk worth investigating	0	No
		The Feminist Jack the Ripper Tour	0	Yes (N)
		Ghost Bus Tour of London	0	No
	Multiple (incl. Whitechapel)	Murder Mile Walks – Guided walk of Soho's most notorious murder cases (12 murderers)	0	No
		The Blood and Tears Walk: Serial killers and other creepy London Horror Stuff (16 murderers)	3 (N)	No
	Ratcliffe Highway Murders	Ratcliff highway Murders (True Crime) Walking Tour	0	No
	None specified	Bleeding Hearts and Body Parts: A London guided walk	0	No
		Medicine, Monks and Murder Walking Tour	0	No
		Ghastly Ghost Walking Tour in London	0	No
		The Cloak and Dagger Tour	0	No
		London Ghost and Infamous Murder Tour	0	No
Brighton	Multiple (all incl. the Trunk Murders)	**Murder Most Foul (unique tour by The Insurance Institute of Sussex)** (2 murderers)	1 (N)	Yes (N)
		The Bad Murder Tour: Brighton After Dark (2 murderers)	1 (P)	No
		Murder, Malice and Mystery: Brighton's free murder tour (3 murderers)	1 (P)	No
Kent	None specified	Rochester Ghost, Murders and Secrets Tour	0	No
		Rotten Ramsgate Tours	0	Yes (N)

(Continued)

Table 1. Continued.

Tour Location	Historical Murderers Featured	Tour Name	Female Murderers Featured: # and Name (N) or Pseudonym (P)	Female Victims Featured: Named (N) or Anonymous (A)
Edinburgh	William Burke & William Hare	Edinburgh's Dark Side Tour	0	No
		Free Ghost Tour	0	No
	Covenanters murdered	Edinburgh Horror Tours	0	No
	James Douglas, 3rd Marquess of Queensbury	Doomed, Dead and Buried	0	No
	None specified	Edinburgh Murder and Mystery Walking Tour	0	No
		The Ladies and Witchery Tour: The Murder and Mystery Tour	0	No
		Hidden and Haunted	0	No
		Gory Stories: The Kids' Tour	0	No
		Ghostly Underground	0	No
Dundee	None specified	Who Dun It? Crimes of Passion Murder Tours	0	No
		Dark Dundee	0	No
Other locations	Harold Shipman; Myra Hindley & Ian Brady	**The Dark Side of Manchester**	1 (N)	No
	James Colder	**Suffolk Walking Festival (Polstead, Suffolk)**	1 (N)	No
	None specified	Spirit of Glasgow, Horror Walk	0	No
		Hangman's Hill Ghost walk: The Original (Epping Forest, Loughton)	0	No
		Chocolate and Death: A tour of Bristol	0	No
		The Bloody Bourton Walking Tour (Burton-on-the-Water, Gloucestershire)	0	No

Note: Tours in bold include featured female figures, either as murderers or victims.

exclusive attention. Another interesting point to note regarding these tours, is that, when the murderer is unknown, the mystery murderers are presumed to be male. Thus, even without a specified gender, the narratives of these tours could be presumed to have masculine gender role-driven undertones. This is visible in the case of 'Jack the Ripper' where it has been suggested that the killer could have been female, but they are generally viewed as having been male and are represented as such in popular culture.

It needs to be noted that both the act of murder and the murderer themselves are socially and legally constructed, through notions of culpability, victim vulnerability, and blame (D'Cruze et al., 2006). According to Davidson and Chesney-Lind (2009, p. 76, emphasis in original), murder is largely understood as a male phenomenon, with women often being considered '*the invisible offender*'. The notion of hegemonic masculinity recognizes the maleness of crime, in general, and of murder, in particular. The emergence of such ideas derives from the social dominance of men wherein there is male agency and female passivity (D'Cruze et al., 2006, p. 40). It has been suggested that male killers might be given greater attention than female killers because they are viewed as a greater threat to society (Gruenewald et al., 2009). However, D'Cruze et al. (2006) argue that women's violence is rationalized within sexist understandings of femininity, womanhood and motherhood as well as the framework of existing beliefs regarding acceptable female behaviour. Thus, due to their deviation from patriarchal gender roles, women who commit violence are 'othered', labelled as monstrous, abnormal, or even as no longer truly a 'woman' (Davidson & Chesney-Lind, 2009; Farrell et al., 2011). For example, when discussing Moors Murderer Myra Hindley, King and Cummins (2017, p. 209) emphasize that 'the female as child killer and the challenge posed to social constrictions of femininity is central to the conceptualisation of Hindley as the most evil woman in Britain or, at best, under [Ian] Brady's spell'.

This stands in stark contrast to male killers who are often presented as compelling, featuring characteristics that appeal to popular audiences and, consequently, are transformed into celebrities to be lauded and worshipped (Oleson, 2006). Such popular discourses relate to accepted narratives of power dynamics within which the male killer is playing out culturally acceptable scripts that form part of modern beliefs (Tithecott, 1997). As such, demonstrations of the male killers' monstrosity both deny the killers' connection with normal life while simultaneously celebrating his 'perceived transcendence of normality' (Tithecott, 1997, p. 9). The result of which is both attractive and repulsive (Oleson, 2006; Tithecott, 1997). Far from the mysterious, enigmatic, talented male killer, violent women are Frankensteinian (Picart, 2006) and so truly monsters. This is clearly visible in the Kriminalmuseum in Vienna, Austria where Huey (2011, p. 390) found that representations of women were either as perpetrators or victims and that 'when they are shown as offenders, their crimes, no matter how pedestrian, are treated as exceptional', which reinforces the gendered stereotype of 'woman as nurturer'. Non-stereotypical homicides, such as those committed by women, receive less cultural attention than those committed by men. While characteristics other than gender are also important, nevertheless men receive greater coverage than women (Gruenewald et al., 2009). Thus, given the general absence of female murderers within the murder tour advertisements, they can be seen as reinforcing this dichotomy wherein men are powerful and have agency in contrast to women who are frail and inactive.

This is particularly evident through the more prevalent presence of the female figure as victim. While it should be noted that there are male victims for some of the male killers included in the tours, many of the victims in these cases are women. Ten of the tour websites mention female murder victims, although only two of these tours are located outside of London. The other eight are focussed on the case of Jack the Ripper. Five of these don't explicitly name the victims with two referring to them merely as 'prostitutes'. While the original Jack the Ripper tour does name all five victims, this information is only located on supplementary pages, not on the homepage (Discovery Tours & Events, 2020b). In contrast, and perhaps unsurprisingly given the tour's name, the Feminist Jack the Ripper Tour not only identifies all five victims but also promotes itself as unique in that it focuses on the lives of the victims, bringing them 'centre stage' (Look Up London Tours, n.d.). However, this tour was noted as being an outlier as the vast majority of tours fail to mention the female victims at all, and those that do often describe them in pejorative terms or use them as theatrical props.

Whereas the serial killer figure is inscribed with power, their victims, often women, especially sex workers, are viewed as weak and lacking cultural presence (Tithecott, 1997). Davidson and Chesney-Lind (2009) note that this lack of power is what leaves women vulnerable to violence and male domination. The prevalence of violence against women then is a manifestation of gender inequality (Garcia-Moreno et al., 2005). Given that this is a serious contemporary issue, the general depersonalization of the victims can be viewed as especially problematic, particularly through the use of the terrorization of women at the hands of men as entertainment often masquerading as education. For example, one tour (Murder Mile Walks) advertises itself with photographs of clients gleefully pretending to strangle each other. When discussing the Jack the Ripper walks, Wilbert and Hansen (2009, p. 201) highlight the problematic nature of their narratives which 'ignore issues of gender politics by fetishizing and trivializing male violence towards women.' Similarly, Huey (2009) noted that, at the Kriminalmuseum in Vienna, there is a specific portion of the museum which focuses on sex and violence wherein images of dead and mutilated female forms are displayed without context, resulting in the fetishization of sexual violence against women. The murder tours then can be seen as entrenching traditional gender roles, where women are passive, powerless victims in the face of male agency and sexual violence. They depict men as those who act upon others, and women as those who have things done to them. This reinforcement of gender roles is then compounded by treating female murders as both extraordinary and gender norm transgressive while also minimizing their narratives within these dark tourism activities.

Conclusions

This work has outlined the issues around the gendered presentation of murder within the context of dark tourism. As has been noted, the relative rarity of female killers, especially serial killers, has created challenges in understanding them. This has been further compounded by restrictive definitions that serve to exclude female serial killers in favour of male ones. Given the academic and criminological association between men and murder, it may be unsurprising that women are often overlooked in murder-related dark tourism activities. However, as has been observed, there have been noted shifts towards the greater inclusion of female killers within the literature (Farrell et al., 2011), and the historical record has a substantial, if not significantly large, pool of potential female subjects for murder tours. Therefore, the absence of female perpetrators within the narratives of existing murder tours potentially points to a separate issue wherein women are often confined to the role of faceless victim, a casualty of male sexual violence. Given this decontextualization and fetishization of female victims as well as the paucity, when compared to their male counterparts, of female killer narratives in murder tours, it can be observed that murder tours may be transgressive in their apparent memorialization of killers, but they are, in many ways, conforming to societal expectations of gendered behaviour, reinforcing narratives of what constitutes traditional male and female roles.

Additionally, while it was not the purpose of this work to dispute the terminology 'dark tourism', it would appear that tourism to sites of crime, particularly violent transgressions, may differ from visits to sites of memorialized tragedy, such as Auschwitz in Poland or Ground Zero in New York. These sites function as a tool for the remembrance of the victims of these events. In contrast, violent criminal tourism activities memorialize the perpetrator, often in an almost celebratory manner, while generally de-humanizing the victims. Given the very clear narrative difference, as well as the difficulty around the use of the term dark tourism in a non-English speaking context (Hartmann, 2014), it is proposed that visitation to these criminal murder sites may be better labelled 'transgression tourism'. This removes the dark–light dichotomy (Seaton, 2009) and instead situates these activities within specific socio-cultural and legal contexts. Afterall, what is transgressive for one culture may not be for another. Furthermore, it highlights a tourism environment which covers both transgressive acts and is in itself transgressive through its celebration of the perpetrators.

It should be noted that this work is theoretical and draws conclusions based on marketed dark tourism attractions associated with murder, and it is possible that there are dark tourism activities which are female perpetrator-focused but lack significant marketing. Additionally, the tours that discuss male killers may also present female examples within the actual context of the tour and merely not publicize this aspect. The analysis of the content of these tours is thus recommended as an area of future research. Moreover, while there have been several studies on motivations of dark tourists, the difference in narrative at murder tourism sites may elicit different responses which could validate the distinction between dark tourism and what we have termed transgression tourism. Furthermore, as there are infamous female killers in other parts of the world, it would be relevant to perform a study on the narratives at tourist sites associated with their crimes in order to see if they support the same traditional gender roles as were observed in this work. Similarly, it would be of interest to determine whether or not dark tourism to murder sites occurs outside of the context of the Western, predominantly English-speaking, world, and, if so, if their own cultural gender norms are reinforced at these sites.

Disclosure statement

No potential conflict of interest was reported by the author(s).

ORCID

Bailey Ashton Adie ⓘ http://orcid.org/0000-0002-0309-4511

References

Anon. (1752). *The trial at large of John Swan and Elizabeth Jeffreys, Spinster, for the murder of ther late uncle Mr. Joseph Jeffreys of Walthamstow in Essex…* (2nd ed). C. Corbett.

Ashworth, G. J., & Tunbridge, J. E. (2017). Death camp tourism: Interpretation and management. In G. Hooper & J. J. Lennon (Eds.), *Dark tourism: Practice and Interpretation* (pp. 69–82). Routledge.

Baldwin, F., & Sharpley, R. (2009). Battlefield tourism: Bringing organised violence back to life. In R. Sharpley, & P. R. Stone (Eds.), *The Darker side of travel: The theory and practice of dark tourism* (pp. 186–206). Channel View Publications.

Beamish. (2016). *Mary Ann Cotton's Teapot at Beamish!* Retrieved February 12, 2020, from http://www.beamish.org.uk/news/mary-ann-cottons-teapot-at-beamish/

Bolan, P., & Simone-Charteris, M. (2018). 'Shining a digital light on the dark': Harnessing online media to improve the dark tourism experience. In P. R. Stone, R. Hartmann, T. Seaton, R. Sharpley, & L. White (Eds.), *The Palgrave handbook of dark tourism studies* (pp. 727–746). Palgrave Macmillan.

Dalton, D. (2015). *Dark tourism and crime*. Routledge.

Davidson, J. T., & Chesney-Lind, M. (2009). Gender and crime. In J. M. Miller (Ed.), *21st century criminology: A reference handbook* (pp. 76–84). SAGE Publications Ltd.

D'Cruze, S., Walklate, S., & Pegg, S. (2006). *Murder*. Willan Publishing.

Discovery Tours & Events. (2020a). *Jack the Ripper tour: A walk worth investigating*. Retrieved January 29, 2020, from https://www.jack-the-ripper-tour.com/

Discovery Tours & Events. (2020b). *Quality London Jack the Ripper walks*. Retrieved February 27, 2020, from https://www.jack-the-ripper-tour.com/walk-description/

Farrell, A. L., Keppel, R. D., & Titterington, V. B. (2011). Lethal Ladies: Revisiting what we know about female serial murderers. *Homicide Studies*, 15(3), 228–252. https://doi.org/10.1177/1088767911415938

Farrell, A. L., Keppel, R. D., & Titterington, V. B. (2013). Testing existing classifications of serial murder considering gender: An exploratory analysis of solo female serial murder. *Journal of Investigative Psychology and Offender Profiling*, 10(3), 268–288. https://doi.org/10.1002/jip.1392

Ferrel, J. (2004). Boredom, crime and criminology. *Theoretical Criminology*, 8(3), 287–302. https://doi.org/10.1177/1362480604044610

Fleming, T. (2007). The history of violence: Mega cases of serial murder, self-propelling narratives and reader engagement. *Journal of Justice and Popular Culture*, 14(3), 277–291.

Foley, M., & Lennon, J. J. (1996). JFK and dark tourism: A fascination with assassination. *International Journal of Heritage Studies*, 2(4), 198–211. https://doi.org/10.1080/13527259608722175

Frew, E. (2017). From living memory to social history: Commemoration and interpretation of a contemporary dark event. In G. Hooper & J. J. Lennon (Eds.), *Dark tourism: Practice and Interpretation* (pp. 160–173). Routledge.

Friedrich, M., Stone, P. R., & Rukesha, P. (2018). Dark tourism, difficult heritage, and memorialisation: A case of the Rwandan genocide. In P. R. Stone, R. Hartmann, T. Seaton, R. Sharpley, & L. White (Eds.), *The Palgrave handbook of dark tourism studies* (pp. 261–289). Palgrave Macmillan.

Garcia-Moreno, C., Heise, L., Jansen, H. A. F. M., Ellsberg, M., & Watts, C. (2005). Violence against women. *Science*, *310*(5752), 1282–1283. https://doi.org/10.1126/science.1121400

Gibson, D. C. (2006). The relationship between serial murder and the American tourism industry. *Journal of Travel & Tourism Marketing*, *20*(1), 45–60. https://doi.org/10.1300/J073v20n01_04

Gruenewald, J., Pizarro, J., & Chermak, S. J. (2009). Race, gender, and the newsworthiness of homicide incidents'. *Journal of Criminal Justice*, *37*(3), 262–272. https://doi.org/10.1016/j.jcrimjus.2009.04.006

Haggerty, K., & Ellerbrok, A. (2011). The social study of serial killers: Kevin Haggerty and Ariane Ellerbrok examine the cultural and historical context of serial killing. *Criminal Justice Matters*, *86*(1), 6–7. https://doi.org/10.1080/09627251.2011.646180

Hanna, S. P., Alderman, D. H., & Bright, C. F. (2018). Celebratory landscapes to dark tourism sites? Exploring the design of southern plantation museums. In P. R. Stone, R. Hartmann, T. Seaton, R. Sharpley, & L. White (Eds.), *The Palgrave handbook of dark tourism studies* (pp. 399–421). Palgrave Macmillan.

Harrison, M. A., Murphy, E. A., Ho, L. Y., Bowers, T. G., & Flaherty, C. V. (2015). Female serial killers in the United States: Means, motives, and makings. *The Journal of Forensic Psychiatry and Psychology*, *26*(3), 383–406. https://doi.org/10.1080/14789949.2015.1007516

Hartmann, R. (2014). Dark tourism, thanatourism, and dissonance in heritage tourism management: New directions in contemporary tourism research. *Journal of Heritage Tourism*, *9*(2), 166–182. https://doi.org/10.1080/1743873X.2013.807266

Heidelberg, B. A. W. (2015). Managing ghosts: Exploring local government involvement in dark tourism. *Journal of Heritage Tourism*, *10*(1), 74–90. https://doi.org/10.1080/1743873X.2014.953538

Hellen, F., Lange-Asschenfeldt, C., Ritz-Timme, S., Verhulsdonk, S., & Hartung, B. (2015). How could she? Psychosocial analysis of ten homicide cases committed by women. *Journal of Forensic and Legal Medicine*, *36*, 25–31. https://doi.org/10.1016/j.jflm.2015.08.007

Hitchcock, T., Shoemaker, R., Emsley, C., Howard, S., McLaughlin, J., et al. (2012). *The old Bailey proceedings online, 1674–1913* (Version 7.0). Retrieved February 12, 2020, from www.oldbaileyonline.org

Huey, L. (2011). Crime behind the glass: Exploring the sublime in crime at the Vienna Kriminalmuseum. *Theoretical Criminology*, *15*(4), 381–399. https://doi.org/10.1177/1362480610397416

Hutchinson, L. (2016, November 7). *Dark Angel Mary Ann Cotton: See the County Durham house where she murdered her last victim*. ChronicleLive. Retrieved February 12, 2020, from https://www.chroniclelive.co.uk/news/north-east-news/dark-angel-mary-ann-cotton-12122102

Ivanova, P., & Light, D. (2018). 'It's not that we like death or anything': Exploring the motivations and experiences of visitors to a lighter dark tourism attraction. *Journal of Heritage Tourism*, *13*(4), 356–369. https://doi.org/10.1080/1743873X.2017.1371181

Kang, E.-J., Scott, N., Lee, T. J., & Ballantyne, R. (2012). Benefits of visiting a 'dark tourism' site: The case of Jeju April 3rd Peace Park, Korea. *Tourism Management*, *33*(2), 257–265. https://doi.org/10.1016/j.tourman.2011.03.004

Kim, S., & Butler, G. (2015). Local community perspectives towards dark tourism development: The case of Snowtown, South Australia. *Journal of Tourism and Cultural Change*, *13*(1), 78–89. https://doi.org/10.1080/14766825.2014.918621

King, M., & Cummins, I. (2017). Take me to the Moors: Mediatised murder forty years on – an analysis of Granada TV's See No evil. *International Journal of Social Science and Humanity*, *7*(4), 206–211. https://doi.org/10.18178/ijssh.2017.V7.821

Lennon, J. (2010). Dark tourism and sites of crime. In D. Botterill & T. Jones (Eds.), *Tourism and crime: Key themes* (pp. 215–228). Goodfellow Publishers Ltd.

Lennon, J., & Foley, M. (2000). *Dark tourism: The attraction of death and disaster*. Continuum.

Light, D. (2017). Progress in dark tourism and thanatourism research: An uneasy relationship with heritage tourism. *Tourism Management*, *61*, 275–301. https://doi.org/10.1016/j.tourman.2017.01.011

Look Up London Tours. (n.d.). The Feminist Jack the Ripper tour. Retrieved June 15, 2020, from https://lookup.london/walking-tours/feminist-jack-the-ripper-walk/

Oleson, J. C. (2006). Contemporary demonology: The criminological theories of Hannibal Lecter, part two. *Journal of Criminal Justice and Popular Culture*, *13*(1), 29–49.

Picart, C. J. S. (2006). Crime and the Gothic: Sexualizing serial killers. *Journal of Criminal Justice and Popular Culture*, *13*(1), 1–18.

Powell, R., & Iankova, K. (2016). Dark London: Dimensions and characteristics of dark tourism supply in the UK capital. *Anatolia*, *27*(3), 339–351. https://doi.org/10.1080/13032917.2016.1191764

Rice, A. (2009). Museums, memorials and plantation houses in the black Atlantic: Slavery and the development of dark tourism. In R. Sharpley & P. R. Stone (Eds.), *The darker side of travel: The theory and practice of dark tourism* (pp. 224–246). Channel View Publications.

Seaton, A. V. (1996). Guided by the dark: From thanatopsis to thanatourism. *International Journal of Heritage Studies*, *2*(4), 234–244. https://doi.org/10.1080/13527259608722178

Seaton, T. (2009). Thanatourism and its discontents: An appraisal of a decade's work with some future issues and directions. In T. Jamal & M. Robinson (Eds.), *The SAGE handbook of tourism studies* (pp. 521–542). Sage Publications Ltd.

Soothill, K., Peelo, M., Francis, B., Pearson, J., & Ackerley, E. (2002). Homicide and the media: Identifying the top cases in *The Times*. *The Howard Journal and Crime and Justice, 41*(5), 401–421. https://doi.org/10.1111/1468-2311.00255

Stone, P. R. (2006). A dark tourism spectrum: Towards a typology of death and macabre related tourist sites, attractions and exhibitions. *Tourism: An Interdisciplinary International Journal, 54*(2), 145–160.

Stone, P. R. (2009). 'It's a bloody guide': Fun, fear and a lighter side of dark tourism at The Dungeon visitor attractions, UK. In R. Sharpley & P. R. Stone (Eds.), *The darker side of travel: The theory and practice of dark tourism* (pp. 167–185). Channel View Publications.

Stone, P. R. (2012). Dark tourism and significant other death: Towards a model of mortality mediation. *Annals of Tourism Research, 39*(3), 1565–1587. https://doi.org/10.1016/j.annals.2012.04.007

Stone, P. R. (2013). Dark tourism scholarship: A critical review. *International Journal of Culture, Tourism and Hospitality Research, 7*(3), 307–318. https://doi.org/10.1108/IJCTHR-06-2013-0039

Stone, P. R., & Sharpley, R. (2008). Consuming dark tourism: A thanatological perspective. *Annals of Tourism Research, 35*(2), 574–595. https://doi.org/10.1016/j.annals.2008.02.003

Timothy, D. J. (2018). Sites of suffering, tourism, and the heritage of darkness: Illustrations from the United States. In P. R. Stone, R. Hartmann, T. Seaton, R. Sharpley, & L. White (Eds.), *The Palgrave handbook of dark tourism studies* (pp. 381–398). Palgrave Macmillan.

Tithecott, R. (1997). *Of men and monsters: Jeffrey Dahmer and the construction of the serial killer*. University of Wisconsin Press.

TripAdvisor. (2020). *Jack the Ripper Tour with 'Ripper-Vision' in London*. Retrieved January 29, 2020, from https://www.tripadvisor.co.uk/AttractionProductReview-g186338-d11468721-Jack_the_Ripper_Tour_with_Ripper_Vision_in_London-London_England.html

Vanneste, D., & Winter, C. (2018). First world War battlefield tourism: Journeys out of the dark and into the light. In P. R. Stone, R. Hartmann, T. Seaton, R. Sharpley, & L. White (Eds.), *The Palgrave Handbook of dark tourism studies* (pp. 443–467). Palgrave Macmillan.

White, L. (2013). Marvellous, murderous and macabre Melbourne: Taking a walk on the dark side. In L. White, & E. Frew (Eds.), *Dark tourism and place identity: Managing and interpreting dark places* (pp. 217–235). Routledge.

Wilbert, C., & Hansen, R. (2009). Walks in spectral space: East London crime scene tourism. In A. Jansson, & A. Lagerkvist (Eds.), *Strange spaces: Explorations into mediated obscurity* (pp. 187–204). Ashgate.

Yoshida, K., Bui, H. T., & Lee, T. J. (2016). Does tourism illuminate the darkness of Hiroshima and Nagasaki? *Journal of Destination Marketing & Management, 5*(4), 333–340. https://doi.org/10.1016/j.jdmm.2016.06.003

Commemorative insights: the best of life, in death

Martin MacCarthy ⓘ and Ker Ni Heng Rigney

ABSTRACT
This study explores visitor experience at the National Anzac Centre in Western Australia using multiple qualitative methods. Initially, Nethnography is used to assemble a blend of lived experience and online non-dialogical commentary. Nethnography (an alternative to Netnography) is used here as a mechanism for data grooming. Three data sets inform this study: 500 Trip Advisor comments, 500 Visitors' Book comments and four days of participant observation. The data are then analysed using Reflexive Thematic Analysis and Leximancer in the unsupervised mode. This methodological collage is designed to improve the veracity of interpretation through both lived experience and triangulation across data sources. Findings suggest a significant visitor-thirst for the positive aspects of commemoration. By the same token no respondent reported being motivated by schadenfreude, mortality salience or death. If a certain fascination with, and commodification of death defines popular dark tourism then commemorative tourism's relegation of death indicates exception. It would seem *commemorists* relegate death and darkness to mere context, while gravitas, ritual and cultural validation transcend the superficial and the kitsch. Meanwhile, visitors to the National Anzac Centre concentrate on more endearing traits including sacrifice, love, loss and the nobility of caritas.

Introduction

I have seen men hard pressed not to weep when their horses were killed. ('Stout Hearts that Never Failed' by Ion Idriess, 1932)

Endearing qualities of the human spirit can be a significant motivator in commemorative pilgrimage. One such pilgrimage involves a journey to the southernmost tip of Western Australia. In this quiet part of the world shady eucalypts line both sides of a road leading up to the Desert Corps Memorial atop Mt Clarence, Albany. Established in 1955 the road is called the Avenue of Honour. This place commemorates Australians who died in service or were killed in action, in all wars. For each tree, there is an interpretive plaque describing in almost mundane monologue the chaos by those who endured the darkness. On one plaque is a tribute by Ion Idriess, a veteran of the Australian 5th Light Horse. It is an ode to the faithful horse; the 'stout hearts that never failed' (Idriess, 1932). More specifically, the 6100 horses who accompanied 41,948 Australian and New Zealand soldiers on their way to Europe from this place – and subsequently the Anzac[1] legend was born.

The use of the horse; the shared trust and emotional bond between animal and carer is a percipient theme used by the nearby National Anzac Centre's curators. This gentle and dignified relationship is emphasised in curatorial effort; specifically, words, prose, artwork and imagery.

Indeed, this deliberate celebration of animal-human friendship is consistent with Ballantyne's challenge to heritage curators to take a positive stance and to use 'their craft to address society's needs' (1998, p. 2; for positive emotions in tourism see also Mitas et al., 2012; Packer et al., 2019. For associated ethics see Timothy, 2011). If we look beyond these relationships and superficial bravado therein lies glimpses of authentic, loving care or [secular] caritas, made all the more poignant during extraordinary and savage times. Such loyalty and loss of innocence in war, both in animals and humans underpins visitor-experience to this place of commemoration and qualifies as the premise of this study.

Aim and contribution

The aim of this study is to investigate the National Anzac Centre with a view to understanding the degree of engagement, associated emotions and typical behaviours associated with Antipodean commemorism. Opened in 2014 the National Anzac Centre and environs hold significant importance to both Australia and New Zealand as this is where troops heading to fight the Great War 1914–1918 assembled for the long journey to an uncertain fate.

A review of key concepts; namely pilgrimage, commemoration and the Antipodean ideals of the Anzac follow. The findings comprise a number of themes, of which three are discussed in detail. This addition to our understanding of visitor motives, in turn, informs stakeholders as to how their efforts are being received and thus provides some insight as to how better manage the perceptual positioning of memorials (see also Dunkley et al., 2011; Poria et al., 2006; Tunbridge & Ashworth, 1996; White & Frew, 2013; Winter, 2009, 2010). For the purposes of this study, commemorism is defined as the portmanteau of tourism with commemoration and associated pilgrimage; commemorism being the singular noun whereas commemorist is the countable noun.

Commemorative pilgrimage

In 2011 Hyde and Harman define pilgrimage as 'a journey to a non-substitutable site embodying the highly valued, the deeply meaningful, or a source of core identity for the traveller' (p. 1343; Lockstone-Binney et al., 2013). They also claim secular pilgrimage has increasingly replaced religious pilgrimage and is important to the 'core identity' of the tourist. Their study of seven visitor groups to the Gallipoli peninsular lists the motives for secular pilgrimage: namely (1) spiritual, (2) nationalistic, (3) family, (4) friendship, and (5) travel (see also Butler & Suntikul, 2018). Brown adds nuance here with her 2016 study of visitors to the graves of Jean Paul Sartre and Simon De Beauvoir in Paris. Brown distils her observations to three motives: (6) a desire for closeness; (7) a wish to pay respects; and (8) an appreciation of the influence the two writers have had on our lives. While it might be a stretch to generalise that all types of secular pilgrimage share these eight motives one must acknowledge Hyde and Harman's data came from Gallipoli visitation and therefore, resonates with commemoration whereas Brown's does not.

Gallipoli refers to the Gallipoli or Dardanelles campaign (Antipodean nomenclature), also known by the Turks or Türkleri as the Battle of Canakkale on the Gelibolu (Gallipoli) peninsular in southern Turkey. During WWI, Australian and New Zealanders identified as members of the British Empire; themselves members of the Entente powers. This coalition was at the time intent on weakening the Ottoman Empire and chose to do so by seising control of the adjacent waterway which was facilitating trade with Russia. The place chosen where the Antipodeans would contribute to this strategy was the Gallipoli peninsular. After several months of costly battle the Entente powers withdrew, essentially admitting defeat.

Although Gallipoli was not considered an ANZAC victory in a military sense the campaign has nevertheless evolved into a seminal commemorative site due to embodying key trans-national Antipodean traits and semiotics. The term Anzac and the date of the landings, 25 April is often used to punctuate the birth of Australia's and New Zealand's independent national identity (Cakar, 2018,

2019; Cheal & Griffin, 2013; Hall et al., 2010; Hyde & Harman, 2011; McKenna & Ward, 2007; Packer et al., 2019; Polonsky et al., 2013; Scates, 2008; Slade, 2003; West, 2010). While over 1000 Anzac memorials and cenotaphs are seeded throughout Australia and New Zealand four main sites commemorating the ANZAC stand out on social media: Gallipoli; The Australian War Memorial, Canberra; The Sir John Monash Centre, France; and the National Anzac Centre (NAC). The NAC in Albany, Western Australia is the focus on this study (see also Inglis & Brazier, 1998; Lee, 2010; Nelson, 2020. See also Ekins, 2010) (Figure 1).

In highlighting the military heritage of a place, one notes the distinction between a cenotaph and a memorial. A cenotaph is an empty tomb designed to commemorate an individual or group of people whose remains lie elsewhere. A memorial is a symbolic structure designed as a focal point for symbolism and not necessarily the identified, or even unidentified dead. The NAC is a somewhat unusual memorial in that it is situated 10,000 km's from the WWI battlefields, and some wonder why it should be so significant (Scates, 2008, 2009; Stephens, 2014). While the journey, war and associated tribulations form the content the curators promote its location as important. The rationale being Antipodean forces concentrated near the NAC prior to departing in two convoys for Europe. From an affective perspective, Albany was the last of Australia thousands of young men and women ever saw. Such poignancy is curatorially highlighted through the story of Faye Howe, the daughter of the lighthouse keeper on nearby Breaksea Island. As the convoys departed it is reported Ms Howe waved at the ships as they passed close by the island. This in turn encouraged the men to write to her from Europe. Such was the volume of letters received the Australian Government donated 4000 one-penny stamps to facilitate her replies, in the hope of lifting the morale of the troops. The story is the subject of a fictional children's book by Dianne Wolfer titled, *Lighthouse Girl* (2010). Later, this inspired the Western Australian Black Swan Theatre's adaptation of the book in 2017 (Turner, 2017). For the importance of stories in

Figure 1. The National Anzac Centre, Albany, Western Australia. © City of Albany. Note: A visual overview of the NAC can be found here: https://youtu.be/yL4qF5LF1BA.

commemoration see Ryan's 'Battlefield tourism: History, place and interpretation' (2007; see also Laing & Frost, 2019).

Commemorative politics

While Hyde and Harman's (2011) motives for secular pilgrimage lends itself to a more pragmatic touristic paradigm, Winter takes a different view of commemoration (2019). Winter's interpretation is more spiritual, albeit secular, focussing on the rituals associated with remembrance (see also Pretes, 2003; Rook, 1985). According to Winter, remembrance and pilgrimage are similarly defined and related to Halbwach's (1992) theory of collective social memory, the essence of which being that pilgrims travel to non-substitutable sites with the aim of connecting with a meaningful past. This is achieved through behaviours such as individual and group rituals; including votive deposition (Hyde & Harman, 2011; Raj & Griffin, 2015; Willson et al., 2013; Winter, 2019) and 'national reflection' (Kennell et al., 2018: Frost & Laing, 2013).

In keeping with Halbwach's collective memory theory is Irvine's (2018) interpretation as one of social memory kept alive by negotiation and social agitation. Irvine notes the waxing and waning over time of pilgrim interest in a site with the catalyst being not the state, or even existence of the physical destination but the 'cultivation of dissonance' related to its meaning (p. 366). Citing the case of pilgrims visiting the site of the former statue of Our Lady of Ipswich (destroyed during the Reformation), Irvine purports that for a place to hold significance and draw devotees it does not have to exist in its original form. Memory lives on regardless and is maintained by a state of social interest, fuelled by agonistic discourse. This idea is relevant to the NAC given it acknowledges but subordinates the importance of physical location in the showcasing of national collective memory. In another sense, this transient nature of collective memory relegates those who are critical of the decision to locate the National Anzac Centre in a place where no conflict occurred, namely Albany (Scates, 2008, 2009; Stephens, 2014).

Location and collective social memory are popular themes encountered in commemorative studies. In Tourism, Fathi, focusses on social discourse and national agenda as the reasons behind Antipodean pilgrimage to the village of Villers Bretonneux in France (2019). Using a somewhat cynical paradigm Fathi claims, 'Commemorative diplomacy cares little about history, but does much to facilitate the country's political and commercial agendas of the day' (para. 33). This is claimed in the context that 'the vast majority of the French are unaware of the Anzacs' [*sic*] (para. 28), but [presumably] appear comfortable with Australia's $104M investment that is the Sir John Monash Centre near Villers Brettoneux. Fathi's perception contrasts markedly with the Centre's Deputy Director, Catherine Carnel. Carnel notes the French are indeed interested in Antipodean involvement during the Great War and particularly the battles fought in the Somme valley. Carnel notes that this interest is growing (personal communication, July 11, 2019; see also Bond, 2002; Inglis & Brazier, 1998; Lee, 2010; Nelson, 2020; Todman, 2005).

This nexus of place, poignancy, and collective social meaning is a powerful facilitator of nationalism and national identity (Kennel et al., 2018; Frew & White, 2015). No doubt justifying the annual pilgrimage of Antipodeans to the Gallipoli peninsular to commune (Holt, 1995) with others over an idea they respect and hold dear (Cakar, 2018, 2019, 2020; Cheal & Griffin, 2013; Hall et al., 2010; Hyde & Harman, 2011; McKenna & Ward, 2007; Polonsky et al., 2013; Scates, 2008, 2009; Slade, 2003; West, 2010). Such appreciation would also no doubt justify the Australian Government's allocation of $498M AUD to the 2020 upgrade of the Australian War Memorial in Canberra. An upgrade that has attracted significant controversy with submissions to a parliamentary inquiry warning of excessive veneration. Which begs the question, how much is enough? It also begs the question of who is being targeted and what this 'therapeutic mileau' will actually do ('Former war memorial', 2020; see also Fathi, 2019; Laing & Frost, 2019). One possibility is attracting tourists for the sake of numbers risking turning commemoration into a theme park experience. As Daley cautions, 'We demean our history when we turn the Australian War Memorial into Disneyland'

(2019). Yet surely extraordinary and emotional moments in a nation's history are deserving of legacy. Hirsch expands further that if memories are traumatic enough that they can be passed on to the next generation in a form of 'postmemory' (2008, 2012; see also Weissman, 2018). And so, the politicising of commemoration continues from one generation to the next.

Method

This study is part of a larger project using data from the National Anzac Centre. The project includes a discussion of Nethnography, which is the marrying of non-dialogical Netnography with lived experience and in doing so qualify as an ethnographic method. The project also includes the introduction of Dialectic Thematic Analysis (DTA), a modular analytical method designed to straddle and bind existing knowledge with new knowledge. This is achieved through the coincidental validation of past published work during the revelation of new knowledge. In actively using past knowledge to inform new knowledge DTA strengthens the link between what is discovered, with what has been purported to exist. The method is dialectical (meaning opposing with a common goal) in the sense simultaneous paths of deduction and induction inform each other, and the findings. This article showcases the inductive path of DTA. The methods highlighted in this study include Nethnography (MacCarthy & Fanning, 2020), Reflexive Thematic Analysis (Braun & Clarke, 2006, 2019), and Content Analysis via Leximancer (Angus, 2014; Smith & York, 2016).

As is typical of induction, themes emerge from a large volume of unstructured data. These themes are essentially grounded in the data although this method is not strictly Grounded Theory. The methodology is nested in interpretivism which is detailed in two seminal works: Lincoln and Guba's (1985) Naturalistic Inquiry, and Denzin and Lincoln's Handbook of Qualitative Research (2000; see also Denzin & Lincoln, 1998; Silverman, 2011). The particular data grooming method used here is a recent adaptation involving the use of Nethnography preceded by a period of familiarisation with the phenomenon; in this case Participant Observation. The combination of lived experience with passive online scraping is referred to as Nethnography[2] For the avoidance of doubt Nethnography is a distinct method to the more traditional dialogical Netnography (see Costello et al., 2017; Kozinets, 1998, 2002, 2010, 2015; MacCarthy & Fanning, 2020).

In this case, two researchers visited the NAC on two occasions in 2018, three months apart for two days each time. Immersed as visitors, we observed and recorded activity while interacting with both the public and custodians. As participants we engaged in conversation, recording comments from 21 visitors and 8 staff while guided by a semi-structured questionnaire (Appendix). More visitors were observed but not engaged. This included inside the NAC where was deemed sensitively inappropriate.

Photo elicitation, and introspection (Wallendorf & Brucks, 1993) informed the process as the authors later reviewed hand-written notes, photographs and video footage of the interior and discussed happenings in the evenings. These data were then compared with a second data set; that being a post-experience corpus. The corpus comprises 500 entries (3337 words) from the NAC visitors' book (see Winter, 2011), with a further 500 NAC-specific comments (31,733 words) compiled from the popular touristic site Trip Advisor (see Cakar, 2018). The scrape was limited to English language speakers. Commentary also includes 533 photographs taken by visitors and attached to their reviews (Timothy & Groves, 2001). This does not include photographs or video taken by the researchers.

The combined digital data set of $n = 1000$ entries are the most recent and therefore considered to be a random proxy. Random in the sense that no comments from a larger available corpus were selected for inclusion, and none from the most recent 500 were excluded. They are simply 500 of the most recent social media, and 500 most recent visitors' book comments. The 1000 comments were then transposed into an Excel spreadsheet where they have been read multiple times. As each potential theme was explored and discussed individual cells of comments were highlighted using Stottok, Bergaus and Gorra's technique of colour-codes and worksheets (2011). The Sort, Search and formula functions of Excel were then used to assist analysis of the data (e.g. LOOKUP, SORT and COUNTIF) (Bree & Gallagher, 2016; Solveig, 2016).

In parallel, both digital data sets were also analysed using the social research software Leximancer. Various analytical aspects of the software were used to explore sentiment, identify co-occurrence, and distil themes. The software's ability to explore relationships was less relevant as the digital data is non-dialogical, meaning there is no interaction between respondents or researchers. Respondents are not engaged in conventional online discourse but merely publishing a single comment in the expectation someone will read and appreciate their effort. With Leximancer, the most useful aspect was the unsupervised (automated) ontological model highlighting occurrence and co-occurrence. The summative aspect of the programme was also helpful. In a sense, computer analytics contemporaneously fortifies researcher observation. To improve the credibility of findings included in the method was a member-check with two drafts forwarded to the staff of the NAC and the City of Albany for comment. No amendments were requested (Denzin & Lincoln, 2000; Wallendorf & Belk, 1989).

Findings

Complementing each other are the two methods of analysis used in this study; namely Reflexive Thematic Analysis (RTA) and computer-assisted Content Analysis (CA). Figure 2 displays the inductive process of RTA, while Figure 3 displays the results of Leximancer. Unlike Leximancer's CA, the inductive process in RTA is overt and can be followed/validated, which in turn encourages credibility. Leximancer's process, however, is latent, proprietary, and couched in the inviolate nature of implied veracity.

Initially, four general groups were used to code the data and within each are related concepts (see Figure 2). These were then combined to produce categories and finally distilled to three overarching themes: *Visitor Experience, Self-development* and *Community Significance*. Given Self-development and Community already appear in the literature (Cakar, 2018, 2019, 2020; Cheal & Griffin, 2013; Fathi, 2019; Hall et al., 2010; Hyde & Harman, 2011; Lockstone-Binney et al., 2013; McKenna & Ward, 2007; Polonsky et al., 2013; Scates, 2008, 2009; Slade, 2003; West, 2010) a focus on the three unique sub themes of *Visitor Experience*: namely *Hero Worship*,

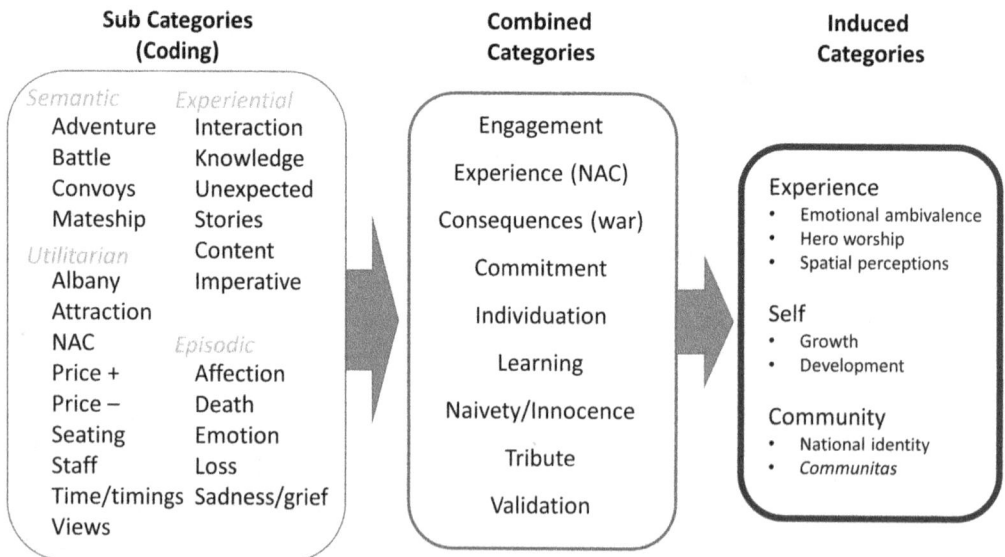

Figure 2. Inductive process.

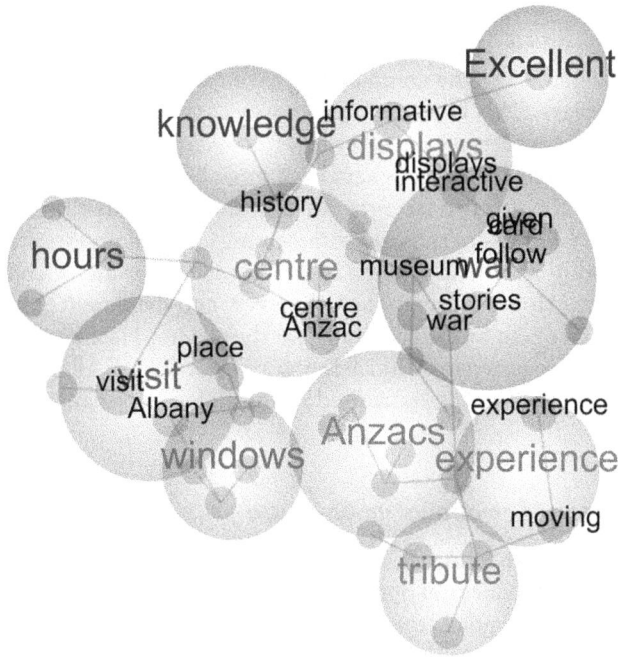

Figure 3. Unsupervised analysis of visitors' book and tip advisor comments.

Emotional Ambivalence, and Spatial Perceptions was deemed more ampliative for this study (for *Communitas* see Celsi et al., 1993).

The three experiential themes are: (1) Emotional ambivalence, referring to the opposing extremes of life and death, while specifically valorising caritas. This in turn encourages respondent empathy. (2) The absence of hero worship or hagiography, which in turn encourages respondent engagement. (3) Expanding the use of physical space, by using digital technology and seamless connections with meaningful environs. This considered use of digital technology encourages perceptions that there is more commemorative space than literal. Prior to discussing these three themes an overview of Leximancer's analysis complements the process.

Figure 3 displays Leximancer's version of the data, teased from an automatic run of the two digitised data sets; the Visitors' Book, and Trip Advisor comments. CSV files of both cohorts were combined and run in an 'unsupervised' [inductive] manner (Angus, 2014; Smith & York, 2016). The default setting of two sentences per considered block was adjusted to accommodate the many comments that were no more than one sentence. 96% of the Visitors' Book have single-sentence comments only whereas the Trip Advisor corpus has an average number of sentences per respondent of 3.9. A point of clarity here; many respondents failed to properly punctuate their comments requiring the data to be 'cleaned' prior to analysis. This includes acronyms, spelling, grammar and syntax errors. The tenet underpinning this process was to keep as much of the original meaning intact.

Analysis

On consideration of Leximancer's concept map (Figure 3) one sees 'War', 'Displays', 'Centre' and 'Anzac' conflated by co-occurrence. This means Anzac is often mentioned in the same sentence as 'Display' and 'Centre'. What is also associated with this nexus is 'Excellent' and related synonyms. This suggests an overwhelmingly positive experience for most respondents. Indeed, only 32 or the 1000 comments contain negative sentiment. The main concern appears to be the price of entry. The price of admissions in the case of detractors is perceived to be high, along with questions being asked as to why it is being charged at all. This implies overt commodification of commemoration is being perceived by respondents to be incongruous (Stone et al., 2018).

Another theme of interest are the commanding views which include the natural harbours of Shoal and Frenchman Bays. This is where both convoys assembled in 1914. The Centre resides atop Mt Clarence in the wider Albany Heritage Park, commanding sweeping views of the approaches to Albany and environs. Designers have capitalised on these views by including four expansive windows in the Centre; two facing south-west and two south-east. Visitors were observed gathering at the windows while admiring and commenting on the uninterrupted views. The theme, 'Windows' co-occurs with 'Visit' and 'Hours' suggesting the views are of communal significance while visiting the Centre. 'Hours' refers to the time taken, and the time required to do the activity justice, or at least the time spent by the respondent at the Centre. The literal time most often mentioned is one to two hours required to adequately absorb the material. Allied with this are the many respondents admitting to, and/or being surprised at how much information is available to warrant this time.

Theme one: the paradox of emotional ambivalence

Two things are clear when considering the nature of the NAC experience; that it strives to make the experience personal and that it does not opine the topic. To clarify, the NAC does not either overtly promote or demote military conflict. Instead, values such a courage, mateship, honour, loyalty, innocence; in many cases couched in endearing naivety are emphasised in curation. The word 'Caritas' in its secular form describes the complex nexus of military camaraderie and mateship which is magnified during times of conflict – 'Brothers [and sisters] in arms' if you will. In commemoration, caritas' etymology precedes its contemporary medieval religious connotation. Caritas can be considered an amalgam of three ancient Greek concepts of love; that being *storge* (empathy), *philia* (friendship among equals) and *agape* (unconditional love for one's God and [institutional: military] 'family'). While each of the three affinities appear in commemorative display, sometimes they are individuated and sometimes combined. The corollary here is the search for a single descriptor that underpins such uncommon affinity, in a contemporary lexicon that otherwise struggles to do the concept justice. The emphasis in this context is extraordinary, selfless loyalty and affection associated with men, women and animals who often willingly risk and sacrifice their lives for colleagues and country.

The notion of companion animals as not only tools but proxy family is not new (Hirschman, 1994; Kylkilahti et al., 2016), but the willingness to risk death across species goes beyond such pedestrian notions. In a commemorative context, it is more similar to Casbeard and Booth's notion of 'exceptionalism' (2012). A case in point is the recent posthumous award of the PDSA Dickin Medal[3] to the Australian SAS Working Dog, Nordenstamm Joep, call-name *Kuga*. In Afghanistan, August 26, 2011 during contact Kuga received multiple gunshot wounds from a lone enemy combatant. Despite being mortally wounded Kuga was not deterred and wrestled with the gunman until incapacitated. This act reportedly drew fire away from his human colleagues and has been subsequently perceived to be selfless. Kuga suffered both mentally and physically during convalescence, eventually succumbing a year later. His death is officially recorded as 'Died of wounds'.

Whether Kuga consciously risked his life to protect his human companions or whether the behaviour was autonomic is arguably moot. Those who steward Kuga's memorial at Campbell Barracks, Perth profoundly believe the former to be the case (name withheld, [Regimental Sergeant Major – custodian; SASR], personal communication, August 24, 2019). From a social science perspective, we must acknowledge that such perceptions are by default, reality. For when considering the implications of social discourse, it matters not what is real, but what respondents believe is real. 'He's the one who chose to go forward. He's the one who chose to take bullets for both me and my mates' (Sergeant J [name withheld] cited in Waseem, 2018, Para. 12). On receiving the award on behalf of the Special Air Service Regiment, Corporal Mark Donaldson [VC] is quoted, 'He just wouldn't give up on his mates' ('Special Operations Dog', 2018, Para. 10). Punctuating this perceived display of cross-species caritas is the citation awarded in 2018 as a tribute to the memory of Kuga: 'For unstinting bravery and life-saving devotion to his handler and his unit, while on patrol with Operation Slipper in Afghanistan, 2011' (Hayne, 2019; Unsung Heroes – Afghanistan, 2020).

Such commemorative emphasis contrasts markedly with Sharpley and Stone's definition of thanatourism, that being underpinned by mortality salience (2011; Stone & Sharpley, 2008). Nor is the data consistent with a macabre interest in death and suffering, as emphasised in Foley and Lennon's (1996) treatise on death-related tourism; or their 'Dark Tourism' definition (2000). Nor is this related to emerging, alternate views of the topic, such as Podoshen, Venkatesh, Wallin, Andrzejewski, and Jin's view of an emerging dystopian dark tourism founded in an uncertain future (2015). Without doubt, the data are unrelated to Gross' emphasis on trauma and the 'negative sublime' (2006). These findings are a distinction, an exception, for instead, the NAC valorises venerable human traits that emerge in extraordinary times – but most importantly, in spite of death. Visitation to the NAC is demonstrably different from the disrespectful kitsch of popular dark tourism. It has more in common with cultural validation, a sense of national duty (Griffins & Sharpley, 2012), and wonderment at what humans and animals are capable of under duress. Visitor experience at the NAC is typified by profound sadness, quiet dignity, wonderment, and the celebration of all that is impressive about the human spirit. A heightened feeling of emotional ambivalence; of all that is good about life, in the context of death. The contrast between remarkable life associated with contextual death is surely an uncomfortable, yet sanguine confliction.

Illustrating this further, the following interpretive description on display at the NAC accompanies a photograph of Brigadier General Harold Edward Elliot, the day after his battalion's action at Fromelles, 1916: 'An abiding memory was Elliot weeping as he shook hands with the pitifully few survivors'. Then there are visitor comments resonating with this tenet;

> … the horrors of war and the bravery and wonderful spirit of those who fight them. (SM117)

> Very thought provoking and reflective of the bravery, mateship and horrors these Men went through and all that have followed them in the conflicts since … very humbling. (SM118)

Returning to cross-species affection which includes the opening epigraph we observe the role of the horse in the NAC is emphasised. This bond between horse and rider is a poignant contributor to the trans-national Anzac spirit. In deference to this, a life-size copper sculpture of a horse and its rider by local artist Bradley Lucas features prominently on display. The sculpture *ANZAC Spirit* resides in one of the main rooms overlooking the harbour (see Figure 4). Some examples of visitor comments regarding the appreciation of this bond include sympathy typically reserved for humans;

> I was close to tears reading the accounts. What really came out was the deep affection for the horses and how devastated the soldiers were that they had to leave them behind. (SM152).

> I loved the see-through sculpture of a soldier watering his horse from his hat, representing the love the men had for their faithful friends … a moving memorial to those brave men, women and their 4-legged mates. (SM113)

> One thing that got me was all the horses, dragged over there and killed. (M40)

Theme two: hero worship

> Greater fates gain greater rewards. (Heraclitus, circa 400BC)

In contrast to traditional cultural norms, it is fair to say that the NAC is not a collection of hagiographies. Instead, curators have deliberately chosen to avoid hero worship preferring instead a neutral or non-judgemental stance. Exactly how much veneration should curators emphasise is of concern as this is fundamental to the experience. From a stated perspective the NAC makes clear their policy on the matter: 'The experience is known for commemorating the war through

Figure 4. Sculpture, *Horse and Rider* overlooking Albany harbour. The inscription reads: *Sharing the last of my water with my old mate. He deserves a drink as much as I before the charge.* © MacCarthy (2018).

the stories of the ANZACs [*sic*] as opposed to telling its own story via pro-war or anti-war senti-ments' (National ANZAC Centre, 2020). This calculated position is unusual given the degree of hagiography on display in other museums and venues. Hero worship of key community figures is often emphasised: including battles, wars, deaths, resurrections, trials, miracles, sacrifices, jour-neys and tribulations. Indeed, writers such as O'Guinn (1991) would have us believe that societies have always needed heroes. 'Heroes and their associated myths help us make sense of our lives … When heroes and gods are reasoned away, a vacuum of anxiety remains' (p. 103). Deliberately taking an egalitarian stance therefore is arguably culturally counter-intuitive and not without risk that some may take umbrage. The overwhelming feedback suggests otherwise however with visitors reporting empathy with the characters as ordinary people caught up in an adventure with tragic consequences:

> What I liked the most is that they showed pictures of soldiers and officers and what they did. The battles also showed lots of pictures. It helped us to get closer to these heroes. (SM86)

Theme three: expanding the physical space

A third notable observation is the use of both symbolism and digital technology to extend the finite physical space. The NAC has co-opted the digital space as an integral part of the experience. Specifi-cally, the use of 9×6.5 cm 'baseball' cards of 32 personas involved with the convoy and subsequent conflict (Figure 5). On the front is a picture of the luminary with their name, role and affiliation. On the back is an optical character recognition symbol which is used to access information via a num-ber of readers throughout the NAC. On leaving, visitors are encouraged to retain their baseball card as a memento. Allied with the use of this technology is the use of touch screens and the use of audio pens using touch-screen activation technology. In combination, these techniques extend the phys-ical space into the digital world of potentially unlimited space. A digital page complements the 32 baseball cards on the main website titled, 'Research an ANZAC' (National Anzac Centre, 2020). Respondent surprise at spatial perception was a common theme:

Figure 5. A sample of the 32 luminaries chosen to adorn NAC interpretive cards. While some are highly decorated, all are portrayed as ordinary people coping in extraordinary circumstances. © City of Albany, 2020.

… it looks small, but there is A LOT of information! (SM192)

I thought it was a lot bigger than it was, but still very impressed. (SM370)

Another technique used to extend the physical space is through linking the perimeter of the NAC with meaningful surroundings. The architecture of the NAC has four large windows by which the Albany's King George Sound features prominently. It is here the two convoys assembled to embark the main portion of the 41,000 troops. In one room there is a large photograph of the sound and a convoy at anchor. Visitors are encouraged to look out at the harbour and imagine what it looked like in 1914.

Extending physical space through co-opting meaningful surroundings is not without precedent. A second room contains an infinity pool pointing towards the passage between Breaksea Island and Torndirrup Peninsula where the convoys left for an uncertain future. A larger infinity pool similarly resides adjacent to the Visitors Centre at the American Cemetery in Colleville sur Mer, France. Both the NAC and the American Cemetery use infinity pools and windows to extend the finite physical space into wider meaningful surroundings. Visitors can then appreciate the importance of location or place and in doing so enhance the experience – by elminating physical boundaries, the experience is less finite.

Conclusions

This study examines attendance and behaviours at the national Anzac Centre in Albany Western Australia. Three sources of data: participant observation, Visitors' Book and social media commentary were compiled using Nethnography which was then analysed using both computer-assisted CA and Reflexive TA. These two qualitative analytical techniques represent the extremes of the Big Q/ little q continuum (Braun & Clarke, 2019; Kidder & Fine, 1987). Resulting themes include: *Experience*, *Self [development]* and *Community*. *Experience* was then parsed with three sub-themes discussed in detail; they are Emotional Ambivalence, Hero worship and Spatial perceptions.

Allied with this are inductive themes teased from the data by Leximancer. They include the co-occurrence of 'War', 'Displays' and 'Centre'. The co-occurrence of 'Anzacs', 'War', 'Experience' and 'Tribute'; and the co-occurrence of 'Visit' with 'Windows' (views) and 'Hours' (time). The co-occurrence of 'Displays' with 'Excellent' and related synonyms suggests a majority appreciation of the experience. The final distillation reveals two overarching Leximancer 'Tags': that being 'War' and

'Visit'. One point of note is the distinction between textual and contextual findings. The study's RTA findings derived from manual coding and informed by the lived experience are more contextual. The lived experience includes participant observation in situ but also a lifetime of prior commemorative experience; both authors have affiliation with the Australian Defence Force. The first author is a former serving soldier and commissioned officer. Indeed, contextual consideration at every step underpins the reflexive nature of RTA. By contrast, Leximancer's findings are entirely non-contextual, metric-based and more akin to CA. The outcome appears more superficial and based on the frequency and co-occurrence of words alone. These words require post analysis contextualising, placing the onus on the researcher to grasp the significance at the end of the latent process. The advantage of computer assisted CA is however that it is less effortful and markedly quicker. It is also a point of note that Leximancer's findings comprise the two digitised cohorts only. The ontological map does not include the lived experience.

By using two complementary analytical methods, RTA and CA one can argue that the findings are more robust. Both methods have their advantages and disadvantages when used independently however when combined there appears to be a synergistic effect regarding insight – they inform each other. Depending on the volume and veracity of CA this has implications for generalisability, or the postmodern equivalent, transferability (Denzin & Lincoln, 1998; Wallendorf & Belk, 1989). The importance of this study lies not only in examining the experience of the NAC for its own sake but also adding, or at least emphasising the notion of positive human traits including caritas as an important, if not an integral criterion of commemoration. A preoccupation with 'darkness' and all the pejorative aspects of conflict has caused us to overlook the gallant deeds and uplifting sentiment that is often associated with war heritage. Such sanguinity is critical to the experience. For many respondents, immersive caritas is their fundamental expectation of commemoration. Judging by the literature the relegation of dark autotelic motives appears to resonate in similar places of commemoration (Brown, 2016; MacCarthy & Willson, 2015; Winter, 2010).

One overarching observation pervading the NAC is that it is not an ostentatious Legenda Aurea[4]-style of repository. It is not replete with heroes deified, idealised and idolised as moral exemplars. Respondents describe a place of significance and cultural identity, but without celebrity heroes. Emerging from the dark, Anzac commemoration is a respectful celebration of all that is potentially good about human dignity, regardless of geopolitics and in spite of war.

As this project seeks to expand our understanding of dark tourism there remains unanswered questions. Questions such as; What is the relationship between hero-worship and visitor empathy. What theories can be incorporated into the architecture and content of heritage sites to enhance visitor expectations of immersive caritas. Is commemoration really dark tourism or should we redefine the hypernym. In Stone's 'Dark to Light' scale, perhaps commemorism prefers the light (2006).

While death underpins both commemorative and dark tourism the attitude to each is antithetical. It appears only in the initial manual coding for RTA and does not appear at all in computer-generated CA. While recreational tourists and 'passers-by' (Brown, 2016) might view death as an autotelic motivator, commemorists relegate death to an instrumental catalyst. The focus at the NAC is not on death itself, but what death facilitates. One conclusion being that motives for visiting such a place appear more related to revelling in what death facilitated in the zeitgeist. Death is, therefore, a catalyst for more positive traits: secular care, overcoming adversity, heroism, courage, and love – love of fraternity (including animals) and love of nation. Visitors are wanting to empathise with the stories and similarly we see empathy contributing to personal and collective identity (Kutbay & Aykac, 2016; Laing & Frost, 2019). A place acknowledged as a 'conduit to reflecting about oneself and others' (p. 197). Hubbert considers grief and crisis from a different perspective, 'It's often loss, it's often crisis, it's often disappointment that has much more to teach us than the bright, shiny moments in life' (Scopelianos, 2020, para 17). Hubbert suggests that grief and crisis are part of everyday life and dealing with this can be a catalyst for wellbeing. Wellbeing defined as not simply happiness but also 'interest, engagement, confidence and affection', along with longer term personal realisations such as control, potential and a sense of purpose (Huppert, 2009). Tourism can benefit from Psychology's paradigm shift; from a focus on dysfunction and disorder to facilitating wellbeing and positive mental states (Argyle, 1999; Huppert, 2009; Seligman, 2002). The notion that commemorative custodians should endeavour to make visitors happy is not necessarily what they want or perhaps need. No visitors to the NAC expressed happiness and yet the majority expressed satisfaction and similar traits of growth, development and wellbeing.

If visitors seek immersion in positive traits should this be emphasised by custodians for a more satisfying experience. Conversely, a focus on negative commemorative emotion may result in long term adverse consequences. Supporting this is Nawihjn and Fricke's study of 240 visitors to the Neuengamme [concentration camp] memorial (2015). Should we therefore de-emphasise the negative and promote the positive or as Cook refers to the concept, 'counter-narratives' (2016). Should we instead pander to consumer demand for an expected experience (Laing & Frost, 2019; Stone, 2006), and what is our responsibility, if any for lasting impressions – including revisit intention (Dimitrovski et al., 2017; Nawijn & Fricke, 2015).

And what of the commodification that straddles both dark and commemorative tourism; specifically, the commercialisation of pilgrimage and entry to memorial sites. There is evidence in these findings of confliction. While some bemoan and punctuate the juxtaposition of charging an entry fee to a sacred site, others accept the pragmatism that staff need to be paid and the facilities maintained. Still others believe the entry fee is beneath such an emotive and transformative experience. One thing is for certain – where phronesis or practical wisdom is concerned, there is still much to be resolved.

Notes

1. ANZAC [acronym] refers to the Australian and New Zealand Army Corps: formed in Egypt 1915. Later, Anzac [proper noun]; refers to Antipodean involvement in all wars and conflict since 1914.
2. Note the difference in spelling. Netnography advocates dialogical interaction and immersion with online respondents. Nethnography advocates non-dialogical scraping of online [*Big*] data with a compulsory period of phenomena immersion, typical lived experience.
3. The People's Dispensary for Sick Animals (PDSA), Dickin Medal is awarded to Commonwealth animals that have displayed 'outstanding gallantry' in wartime.
4. Legenda Aurea, the original [Latin] title for The Golden Legend; 100+ hagiographies compiled by Jacobus de Varagine (Ca. thirteenth century).

Acknowledgements

The authors wish to thank the curators of the National Anzac Centre and the City of Albany their valuable insight and assistance.

Disclosure statement

No potential conflict of interest was reported by the author(s).

ORCID

Martin MacCarthy http://orcid.org/0000-0001-5226-1024

References

Angus, D. (2014, April 3). *Leximancer tutorial 2014* [Video file]. YouTube. https://www.youtube.com/watch?v=F7MbK2AF0qQ&t=2228s

Argyle, M. (1999). Causes and correlates of happiness. In D. Kahneman, E. Diener, & N. Schwartz (Eds.), *Well-being: The foundations of hedonic psychology* (pp. 353–373). Russell Sage Foundation.

Ballantyne, R., & Uzzell, D. (1998). *Contemporary issues in heritage and environmental interpretation: Problems and prospects.* Stationery Office.

Bond, B. (2002). *The unquiet Western front: Britain's role in literature and history.* Cambridge University Press.

Braun, V., & Clarke, V. (2006). Using thematic analysis in psychology. *Qualitative Research in Psychology, 3*(2), 77–101. https://doi.org/10.1191/1478088706qp063oa

Braun, V., & Clarke, V. (2019). Reflecting on Reflexive Thematic Analysis. *Qualitative Research in Sport, Exercise and Health, 11*(4), 589–597. https://doi.org/10.1080/2159676X.2019.1628806

Bree, R. T., & Gallagher, G. (2016). Using Microsoft Excel to code and thematically analyse qualitative data: A simple, cost-effective approach. *AISHE-J: The All Ireland Journal of Teaching and Learning in Higher Education, 8*(2), 2811–28114.

Brown, L. (2016). Tourism and pilgrimage: Paying homage to literary heroes. *International Journal of Tourism Research, 18*(2), 167–175. https://doi.org/10.1002/jtr.2043

Butler, R., & Suntikul, W. (Eds.). (2018). *Tourism and religion: Issues and implications.* Blue Ridge Summit.

Cakar, K. (2018). Experiences of visitors to Gallipoli, a nostalgia-themed dark tourism destination: An insight from TripAdvisor. *International Journal of Tourism Cities, 4*(1), 98–109. https://doi.org/10.1108/IJTC-03-2017-0018

Cakar, K. (2019). Transnational tourism experiences at Gallipoli. *Tourism Management, 74*, 411–412. https://doi.org/10.1016/j.tourman.2019.04.026

Cakar, K. (2020). Investigation of the motivations and experiences of tourists visiting the Gallipoli peninsula as a dark tourism destination. *European Journal of Tourism Research, 24*, 2405.

Casbeard, R., & Booth, C. (2012). Postmodernity and the exceptionalism of the present in dark tourism. *Journal of Unconventional Parks, Tourism and Recreational Research, 4*(1), 2–8.

Celsi, R. L., Rose, R. L., & Leigh, T. W. (1993). An exploration of high-risk leisure consumption through skydiving. *Journal of Consumer Research, 20*(1), 1–1. https://doi.org/10.1086/209330

Cheal, F., & Griffin, T. (2013). Pilgrims and patriots: Australian tourist experiences at Gallipoli. *International Journal of Culture, Tourism and Hospitality Research, 7*(3), 227–241. https://doi.org/10.1108/IJCTHR-05-2012-0040

Cook, M. R. (2016). Counter-narratives of slavery in the deep south: The politics of empathy along and beyond river road. *Journal of Heritage Tourism, 11*(3), 290–308. https://doi.org/10.1080/1743873X.2015.1100624

Costello, L., McDermott, M. L., & Wallace, R. (2017). Netnography: Range of practices, misperceptions, and missed opportunities. *International Journal of Qualitative Methods, 16*(1), 1–12. https://doi.org/10.1177/1609406917700647

Daley, P. (2019, September 5). We demean our history when we turn the Australian War memorial into Disneyland. *The Guardian.* https://www.theguardian.com/australia-news/postcolonial-blog/2019/sep/05/we-demean-our-history-when-we-turn-the-australian-war-memorial-into-disneyland

Denzin, N. K., & Lincoln, Y. S. (1998). *Strategies of qualitative inquiry.* Sage.

Denzin, N. K., & Lincoln, Y. S. (2000). *Handbook of qualitative research.* Sage.

Dimitrovski, D., Senic, V., Maric, D., & Marinkovic, V. (2017). Commemorative events at destination memorials – a dark (heritage) tourism context. *International Journal of Heritage Studies, 23*(8), 695–708. https://doi.org/10.1080/13527258.2017.1317645

Dunkley, R., Morgan, N., & Westwood, S. (2011). Visiting the trenches: Exploring meanings and motivations in battlefield tourism. *Tourism Management, 32*(4), 860–868. https://doi.org/10.1016/j.tourman.2010.07.011

Ekins, A. (2010). *1918 year of victory: The end of the Great War and the shaping of history* (1st ed.). Exisle Publishing.

Fathi, R. (2019). Do 'the French' care about Anzac? *The Conversation.* https://theconversation.com/friday-essay-do-the-french-care-about-anzac-110880

Former war memorial. (2020, June 16). *The Guardian.* https://www.theguardian.com/australia-news/2020/jun/16/former-war-memorial-heads-join-call-to-redirect-500m-for-grandiose-expansion-to-veterans

Frew, E., & White, L. (2015). Commemorative events and national identity: Commemorating death and disaster in Australia. *Event Management: An International Journal, 19*(4), 509–524.

Frost, W., & Laing, J. (2013). *Commemorative events: Memory, identities, conflict.* Routledge.

Griffins, I., & Sharpley, R. (2012). Influences of nationalism on tourist-host relationships. *Annals of Tourism Research, 39*(4), 2051–2072. https://doi.org/10.1016/j.annals.2012.07.002

Gross, A. (2006). Holocaust tourism in Berlin: Global memory, trauma and the 'negative special operations dog (Oct 26, 2018), sublime'. *Journeys, 7*(2), 73–100. https://doi.org/10.3167/jys.2006.070205

Halbwachs, M. (1992). *On collective memory* (L. Coser, Ed.). University of Chicago Press.

Hall, J., Basarin, J., & Lockstone-Binney, L. (2010). An empirical analysis of attendance at a commemorative event: Anzac Day at Gallipoli. *International Journal of Hospitality Management, 29*(2), 245–253.

Hayne, J. (2019, October 26). 'Victoria cross for animals': Military dog Kuga posthumously awarded Dickin Medal for bravery. https://www.abc.net.au/news/2018-10-26/dickin-medal-awarded-kuga-dog-canberra/10433054

Hirsch, M. (2008). The generation of postmemory. *Poetics Today, 29*(1), 103–128.

Hirsch, M. (2012). *The generation of postmemory: Writing and visual culture after the Holocaust*. Columbia University Press.

Hirschman, E. C. (1994). Consumers and their animal companions. *Journal of Consumer Research, 20*(4), 616–632. https://doi.org/10.1086/209374

Holt, D. B. (1995). How consumers consume: A typology of consumption practices. *Journal of Consumer Research, 22*(1), 1–1. https://doi.org/10.1086/209431

Huppert, F. A. (2009). Psychological well-being: Evidence regarding its causes and consequences. *Applied Psychology: Health and Well-Being, 1*(2), 137–164. https://doi.org/10.1111/j.1758-0854.2009.01008.x

Hyde, K., & Harman, S. (2011). Motives for a secular pilgrimage to the Gallipoli battlefields. *Tourism Management, 32*(6), 1343–1351. https://doi.org/10.1016/j.tourman.2011.01.008

Idriess, I. (1932). *The desert column*. Angus & Robertson Publishers Australia.

Inglis, K., & Brazier, J. (1998). *Sacred places: War memorials in the Australian landscape*. Melbourne University Press.

Irvine, R. (2018). Our Lady of Ipswich: Devotion, dissonance, and the agitation of memory at a forgotten pilgrimage site. *Journal of the Royal Anthropological Institute, 24*(2), 366–384. https://doi.org/10.1111/1467-9655.12815

Kennell, J., Šuligoj, M., & Lesjak, M. (2018). Dark events: Commemoration and collective memory in the former Yugoslavia. *Event Management, 22*(6), 945–963.

Kidder, L. H., & Fine, M. (1987). Qualitative and quantitative methods: When stories converge. *New Directions for Program Evaluation, 1987*(35), 57–75. https://doi.org/10.1002/ev.1459

Kozinets, R. V. (1998). On Netnography: Initial reflections on consumer research investigations of cyberculture. *Advances in Consumer Research, 25*, 366–371.

Kozinets, R. V. (2002). The field behind the screen: Using Netnography for marketing research in online communities. *Journal of Marketing Research, 39*(1), 61–72. https://doi.org/10.1509/jmkr.39.1.61.18935

Kozinets, R. V. (2010). *Doing ethnographic research online*. Sage.

Kozinets, R. V. (2015). *Netnography: Redefined*. Sage.

Kutbay, E. Y., & Aykac, A. (2016). Battlefield tourism at Gallipoli: The revival of collective memory, the construction of national identity and the making of a long-distance tourism network. *Almatourism Journal of Tourism, Culture and Territorial Development, 7*(5), 61–83.

Kylkilahti, E., Syrjala, H., Autio, J., Kuismin, A., & Autio, M. (2016). Understanding co-consumption between consumers and their pets. *International Journal of Consumer Studies, 40*(1), 125–131. https://doi.org/10.1111/ijcs.12230

Laing, J. H., & Frost, W. (2019). Presenting narratives of empathy through dark commemorative exhibitions during the centenary of World War One. *Tourism Management, 74*, 190–199. https://doi.org/10.1016/j.tourman.2019.03.007

Lee, T. (2010, April 24). Set in stone: 'A nation of small-town memorials'. https://www.abc.net.au/news/2010-04-24/set-in-stone-a-nation-of-small-town-memorials/408832

Lennon, J. J., & Foley, M. (1996). JFK and dark tourism: A fascination with assassination. *International Journal of Heritage Studies, 2*(4), 198–211. https://doi.org/10.1080/13527259608722175

Lennon, J. J., & Foley, M. (2000). *Dark tourism*. Cengage.

Lincoln, Y. S., & Guba, E. G. (1985). *Naturalistic inquiry*. Sage.

Lockstone-Binney, L., Hall, J., & Atay, L. (2013). Exploring the conceptual boundaries of diaspora and battlefield tourism: Australians' travel to the Gallipoli battlefield, Turkey, as a case study. *Tourism Analysis, 18*(3), 297–311. https://doi.org/10.3727/108354213X13673398610736

MacCarthy, M. J., & Fanning, S. (2020). From Netnography to *Nethnography*: An Anzac commemorative experience trial. *Tourism Analysis*. https://doi.org/10.3727/108354220X15957939969805.

MacCarthy, M. J., & Willson, G. (2015). The business of D-Day: An exploratory study of consumer behaviour. *International Journal of Heritage Studies, 21*(7), 698–715. https://doi.org/10.1080/13527258.2014.1001423

McKenna, M., & Ward, S. (2007). 'It was really moving, mate': The Gallipoli pilgrimage and sentimental nationalism in Australia. *Australian Historical Studies, 38*(129), 141–151. https://doi.org/10.1080/10314610708601236

Mitas, O., Yarnal, C., & Chick, G. (2012). Jokes build community: Mature tourists' positive emotions. *Annals of Tourism Research, 39*(4), 1884–1905. https://doi.org/10.1016/j.annals.2012.05.003

National ANZAC Centre. (2020). www.nationalanzaccentre.com.au

Nawijn, J., & Fricke, M. C. (2015). Visitor emotions and behavioral intentions: The case of concentration camp memorial Neuengamme. *International Journal of Tourism Research, 17*(3), 221–228. https://doi.org/10.1002/jtr.1977

Nelson, B. (2020). Every town, every memorial. *Places of Pride*. https://placesofpride.awm.gov.au/

O'Guinn, T. C. (1991). Touching greatness: The central Midwest Barry Manilow fan club. In R. W. Belk (Ed.), *Highways and buyways: Naturalistic research from the consumer behavior odyssey* (pp. 102–110). Association for Consumer Research.

Packer, J., Ballantyne, R., & Uzzel, D. (2019). Interpreting war heritage: Impacts of Anzac museum and battlefield visits on Australians' understanding of national identity. *Annals of Tourism Research, 76*, 105–116. https://doi.org/10.1016/j.annals.2019.03.012

Podoshen, J., Venkatesh, V., Wallin, J., Andrzejewski, S., & Jin, Z. (2015). Dystopian dark tourism: An exploratory examination. *Tourism Management, 51*, 316–328. https://doi.org/10.1016/j.tourman.2015.05.002

Polonsky, M., Hall, J., Vieceli, J., Atay, L., Akdemir, A., & Marangoz, M. (2013). Using strategic philanthropy to improve heritage tourist sites on the Gallipoli peninsula, Turkey: Community perceptions of changing quality of life and of the sponsoring organization. *Journal of Sustainable Tourism, 21*(3), 376–376. https://doi.org/10.1080/09669582.2012.699061

Poria, Y., Reichel, A., & Biram, A. (2006). Heritage site management: Motivations and expectations. *Annals of Tourism Research, 33*(1), 162–178. https://doi.org/10.1016/j.annals.2005.08.001

Pretes, M. (2003). Tourism and nationalism. *Annals of Tourism Research, 30*(1), 125–142. https://doi.org/10.1016/S0160-7383(02)00035-X

Raj, R., & Griffin, R. (Eds.). (2015). *Religious tourism and pilgrimage management: An international perspective.* CABI Publishing.

Rook, D. (1985). The ritual dimension of consumption. *Journal of Consumer Research, 12*(3), 251–264. https://doi.org/10.1086/208514

Ryan, C. (2007). Battlefield tourism: History, place and interpretation. In *Advances in tourism research.* Elsevier. https://www-sciencedirect-com.ezproxy.ecu.edu.au/book/9780080453620/battlefield-tourism

Scates, B. (2008). Memorialising Gallipoli: Manufacturing memory at Anzac. *Public History Review, 15*, 60. https://doi.org/10.5130/phrj.v15i0.820

Scates, B. (2009). Manufacturing memory at Gallipoli. In M. Keren, & H. Herwig (Eds.), *War memory and popular culture* (pp. 57–75). McFarland.

Scopelianos, S. (2020, August 23). The pursuit of wellbeing (and why you can have it without happiness). https://www.abc.net.au/news/2020-08-23/coronavirus-stop-chasing-happiness-and-focus-on-wellbeing/12566340

Seligman, M. (2002). *Authentic happiness.* Free Press.

Sharpley, R., & Stone, P. R. (2011). *Tourist experience: Contemporary perspectives.* https://ebookcentral.proquest.com/lib/ecu/reader.action?docID=592952

Silverman, D. (Ed.). (2011). *Qualitative research: Issues of theory, method and practise.* Sage.

Slade, P. (2003). Gallipoli thanatourism: The meaning of Anzac. *Annals of Tourism Research, 30*(4), 779–794. https://doi.org/10.1016/S0160-7383(03)00025-2

Smith, A., & York, S. (2016, July 21). *What's new in Leximancer V4.5* [Video file]. YouTube. https://www.youtube.com/watch?v=FB7KIQSE4-Q

Solveig, O. (2016). Using excel and word to structure qualitative data. *Journal of Applied Social Science, 10*(2), 147–162. https://doi.org/10.1177/1936724416664948

Special operations dog posthumously awarded top bravery honour. (2018, October 26). *Forces Network.* https://www.forces.net/news/special-operations-dog-awarded-top-military-honour

Stephens, J. R. (2014). Sacred landscapes: Albany and Anzac pilgrimage. *Landscape Research, 39*(1), 21–39. https://doi.org/10.1080/01426397.2012.716027

Stone, P., Hartmann, R., Seaton, A., Sharpley, R., & White, L. (Eds.). (2018). *The Palgrave handbook of dark tourism studies* (Palgrave handbooks). Palgrave Macmillan.

Stone, P. R. (2006). A dark tourism spectrum: Towards a typology of death and macabre related tourist sites, attractions and exhibitions. *Tourism: An Interdisciplinary International Journal, 54*(2), 145–160.

Stone, P. R., & Sharpley, R. (2008). Consuming dark tourism: A thanatological perspective. *Annals of Tourism Research, 35*(2), 574–595. https://doi.org/10.1016/j.annals.2008.02.003

Stottok, B. O., Bergaus, M. N., & Gorra, A. (2011, January). Color coding: An alternative to analyse empirical data via grounded theory. *Proceedings on the European Conference on Research Methods, Normandy France* (p. 472). ECRM.

Timothy, D. (2011). *Cultural heritage and tourism: An introduction.* Channel View Publications.

Timothy, D. J., & Groves, D. L. (2001). Research note: Webcam images as potential sources of tourism research. *Tourism Geographies, 3*(4), 394–404. https://doi.org/10.1080/146166800110070487

Todman, D. (2005). *The Great War: Myth and memory.* Hambledon and London.

Tunbridge, J. E., & Ashworth, G. J. (1996). *Dissonant heritage: The management of the past as a resource in conflict.* Wiley.

Turner, H. (2017). *Lighthouse girl: Regional tour* [Performed April 17–May 5, 2018]. Black Swan State Theatre Company of WA.

Unsung heroes – Afghanistan. (2020). *Australian War Memorial.* https://www.awm.gov.au/collection/C2641266

Wallendorf, M., & Belk, R. W. (1989). Assessing trustworthiness in naturalistic consumer research. In E. C. Hirschman (Ed.), *Interpretive consumer research* (pp. 69–84). Association for Consumer Research.

Wallendorf, M., & Brucks, M. (1993). Introspection in consumer research: Implementation and implications. *Journal of Consumer Research, 20*(3), 339–359. https://doi.org/10.1086/209354

Waseem, Z. (2018, October 26). Kuga may have saved the lives of his unit. Now he's being recognised. *Canberra Times*. https://www.canberratimes.com.au/story/6001099/kuga-may-have-saved-the-lives-of-his-unit-now-hes-being-recognised/

Weissman, G. (2018). *Fantasies of witnessing: Postwar efforts to experience the Holocaust*. Cornell University Press. https://doi.org/10.7591/9781501730054

West, B. (2010). Dialogical memorialization, international travel and the public sphere: A cultural sociology of commemoration and tourism at the first world war Gallipoli battlefields. *Tourist Studies*, *10*(3), 209–225. https://doi.org/10.1177/1468797611407756

White, L., & Frew, E. (Eds.). (2013). *Dark tourism and place identity: Managing and interpreting dark places*. Routlege.

Willson, G., McIntosh, A., & Zahra, A. (2013). Tourism and spirituality: A phenomenological analysis. *Annals of Tourism Research*, *42*, 150–168. https://doi.org/10.1016/j.annals.2013.01.016

Winter, C. (2009). Tourism, social memory and the Great War. *Annals of Tourism Research*, *36*(4), 607–626. https://doi.org/10.1016/j.annals.2009.05.002

Winter, C. (2010). Battlefield visitor motivations: Explorations in the Great War town of Ieper, Belgium. *International Journal of Tourism Research*, *13*(2), 164–176. https://doi.org/10.1002/jtr.806

Winter, C. (2011). First World War cemeteries: Insights from visitor books. *Tourism Geographies*, *13*(3), 462–479. https://doi.org/10.1080/14616688.2011.575075

Winter, C. (2019). Pilgrims and votives at war memorials: A vow to remember. *Annals of Tourism Research*, *76*, 117–128. https://doi.org/10.1016/j.annals.2019.03.010

Wolfer, D. (2010). *Lighthouse girl*. Fremantle Arts Press.

Appendix

Participant observation phase, Albany Heritage Park

Semi-structured questions asked in no particular order.

Where are you from?

What are you doing while here?

- How long are you planning to spend doing … ?
- How long did you spend doing … ?
- Are you visiting any other part of Albany or the South West?

How is this experience relevant to you?

Are you connected to anyone from the original Anzacs?

- If yes, tell me about them.

Who are you with today?

- How are they related?
- What is their connection with the NAC?

What are your thoughts on the experience?

- What did you learn?
- How has the visit changed you?
- Did it meet your expectations? If not, what can be improved?
- Were there any interesting connections interactions?

Does emotional engagement matter in dark tourism? Implications drawn from a reflective approach

Marianna Sigala ⓘⒹ and Effie Steriopoulos ⓘⒹ

ABSTRACT

Despite the burgeoning research in dark tourism, there is still no universal acceptance of its definition. Past research is criticised for focusing on the motivation rather than the nature and consequences of the dark tourism experience, specifically their phenomenological and contextual basis. This study contributes to the field by adopting a reflective autoethnographic approach for providing a better understanding of dark tourism experiences. The study critically reflects on the researcher's immersive experiences at three USA dark sites (Ground Zero, Gettysburg and Ellis Island). The findings reveal that emotional engagement (type and intensity of emotions elicited during the dark tourism experience) plays an important role in probing and helping visitors to generate meaning through their dark tourism experiences. The study expands the literature on dark tourism experiences by proposing an adapted dark tourism typology framework whereby emotional engagement is used as an explanatory theoretical concept to better identify and understand the nuanced types of dark tourism experiences. In addition, by adopting a multi-disciplinary and experienced-focused approach, the study also contributes to dark tourism research by providing a theoretical underpinning and practical evidence on the sources and processes helping visitors to generate meaning, which is deemed central to immersive dark experiences.

Introduction

As a phenomenon, dark tourism is traced back to centuries ago, attracting interest since ancient times (Seaton, 1999, 2002). However, despite the burgeoning number of studies in dark tourism during the last decades, there is still no universal agreement as to what constitutes dark tourism, how dark tourism differs with other 'niche' forms of tourism such as, heritage and adventure tourism (e.g. visits to conflict zones or dangerous places, ghost tours, thana-themed tourism parks) (Light, 2017). This is also evident in the plethora of terms used to describe this niche form of tourism (Novelli, 2005), which mainly refer to visits of sites where death has taken place (Foley & Lennon, 1996). In addition to 'dark tourism' (Lennon & Foley, 2000; White & Frew, 2013), many other terms are used to describe this phenomenon such as, 'thanatourism' (Slade, 2003; Seaton, 2002), 'battlefield tourism' (Winter, 2011), heritage tourism' (Biran & Hyde, 2013), and 'secular pilgrimage' (Hall et al., 2018; Hyde & Harman, 2011). Light (2017) summarised many other related terms such as 'suicide tourism', 'atomic tourism' 'grief tourism', 'genocide tourism', 'disaster tourism', 'favela tourism' and 'conflict heritage tourism'. This study adopts the dark tourism definition by Lennon and Foley (2000, p. 11) who conceptualised dark tourism as a phenomenon of 'post modernity' that attracts tourists interested in sites connected to death.

Current developments further contribute to the increasing dilution and complexity of the concept of dark tourism (Sharpley, 2009; Stone et al., 2018). A diversified and sophisticated demand shows that not all 'dark' tourists are interested in death, while dark tourism goes beyond visiting places of death or associated with death to include thanascapes commercialising the theme of death. As a result, dark tourism frameworks have been refined to include the wide diversity of dark places and of visitors' motivations. Several typologies of dark tourism have also emerged featuring different 'shades' of darkness in order to embrace all the various forms and intensity of dark tourism supply and demand (e.g. Stone, 2006; Raine, 2013). Indeed, most of the frameworks aiming to conceptualise dark tourism are either supply or demand focused, and they also use the various shades of darkness to capture the heterogeneity of the dark tourism supply or of the visitors' motivations (Light, 2017). Some other frameworks (e.g. Sharpley, 2005) also tried to integrate both the demand and the supply perspectives by introducing scales of grey to reflect the reality that there are places that are not intended to be 'dark' and that not all dark tourists are interested in death. However, no typology has received a universal acceptance, because typologies miss the point that sites or places are not intrinsically (or objectively) dark (Ashworth & Isaac, 2015). Moreover, 'darkness' cannot be viewed as an objective fact because it is subjectively and socially constructed, since (different) people in various (cultural or social) contexts understand and experience dark tourism in different ways (Light, 2017; Prayag et al., 2017). In addition, similarly to findings from the service and experience research (Jain et al., 2017), dark tourism experiences are co-created (Magee & Gilmore, 2015), a concept which supports the notion of an exchange process between the visitor and the site or the site manager. It is not surprising that the existing dark tourism typologies and frameworks have received a lot of scrutiny, because they attribute the label dark to something in an arbitrary process that also ignores the subjective and contextual ways in which dark tourism experiences are co-created and consumed. In line with the current service literature on co-creation and experiences, this paper argues that dark tourism experiences are subjectively, phenomenologically and contextually determined based on the visitor's profile and situation. Our argument addresses an earlier call of the literature advocating that a focus on experiences can be a better way to research and understand dark tourism, because it integrates its demand and supply perspectives (Johnston, 2013).

The need to re-orient dark tourism research into an experience focus is also advocated by recent conceptualisations of dark tourism highlighting that dark tourism constitutes a particular type of experience based on emotional engagement (Oren et al., 2021). As most people visit dark places from a specific interest in death, then what is important is not only the motivation but the consequences of these visits (Stone, 2012). For example, recently Oren et al. (2021) found that the negative emotions felt by the visitors at one of the darkest sites (i.e. Auschwitz concentration camp in Poland) significantly contributed to their experience, and so, they called for urgent research investigating the relation between visitors' emotional engagement and experience at dark tourism sites. They also argued that research focusing on the role of negative emotions is also warranted in order to challenge the conventional thinking and literature that negative emotions lead to negative customer experiences and dissatisfaction. Prior dark tourism research had generated a new theory of tourism consumption for better conceptualising the phenomenon; the mortality mediation model proposed a new way of thinking about dark tourism by focusing on the visitors' experiences and the implications of these experiences on the visitors (Stone, 2012). The mortality mediation model highlighted the opportunities afforded by dark tourism to elicit particular emotional experiences and engage visitors into a reflection and contemplation of the nature of mortality (Light, 2017). However, because adopting this perspective required an understanding of the changing nature of societal relations with death, dying and the dead (Stone, 2012), researchers avoided adopting this different way of thinking so far. Thus, the mortality mediation theory still awaits fuller investigation and development (Light, 2017). Of particular interest and current research priority is the further investigation of the introspective, sensory, transformative and spiritual dimensions of the experiences that may be triggered by the visitors' emotions and affects generated by dark tourism (Light, 2017; Oren et al., 2021; Zheng et al., 2019). Exploring these issues requires openness to more disciplinary perspectives and lenses for seeing things (Light, 2017). The latter is already accepted, as dark tourism is already noteworthy for its multi-disciplinary nature.

To address these gaps, this study adopts an experience-based perceptive as a more appropriate way for exploring dark tourism that seeds better light into the nature and the impacts of dark tourism experiences on visitors' emotional engagement. By focusing specifically on the emotional and affective dimensions of dark tourism experiences, this study offers a deeper understanding on how dark tourism can influence the visitors' reflective and meaning-making processes to achieve transformative impacts. In addition, following previous suggestions, this study contributes to dark tourism research by adopting a multi-disciplinary approach for theorising the nature and impacts of dark tourism experiences. The study draws on theories from consumer behaviour, psychology, service and tourism research for understanding how emotions and feelings triggered through dark tourism experiences lead the visitors to reflect and (re)-think about death related issues and to engage into meaning-making processes that in turn enable them to (re)-set their understanding of dark places and death.

In responding to Light's (2017) argument that future research in dark tourism experiences requires new methodologies, the study adopted a reflective approach [based on the phenomenological practice by Van Manen (2016)] for collecting and analysing primary data. This approach considers the phenomenological aspect of dark tourism experiences, that previous studies have ignored, while it also reflects the researcher's own responsibility to constantly and personally engage reflectively with his/her particular field of interest (Khoo-Lattimore, 2018). Specifically, the study is based on the personal reflections of the researcher undertaking a journey to three different dark sites in the USA (Ground Zero, Gettysburg and Ellis Island). Engagement with the experience and reflection forms part of a phenomenological enquiry that is applied when exploring phenomena. The researcher reflects on the type and the intensity of emotions elicited during the visits in order to derive a better essence of the dark tourism experiences and their impact on the self. Overall, the findings from the researcher's immersive experiences suggest that the mixed emotions triggered and felt in various intensity levels during the dark tourism experiences better reflect and corroborate the diverse spectrum of darkness that is previously found in the literature (Stone, 2006). Consequently, the study suggests that the emotional engagement of dark visitors can help us to better understand the phenomenon of dark tourism, as it more appropriately captures the nuanced models of dark tourists (Light, 2017). Hence, the study concludes by proposing an adapted dark tourism typology framework (initially developed by Stone, 2006) that uses the emotional engagement of visitors during their dark experiences as a new theoretical lens that better explains the various facets, forms and impacts of dark tourism. The adapted framework highlights the role and the importance of studying and using the visitors' emotional engagement (i.e. the type and the intensity of the felt emotions) as a better way for exploring and contributing to dark tourism research, because it addresses two previously identified research gaps (Light, 2017): (1) it uses a phenomenological, experienced-focused approach for better understanding dark tourism experiences and for theorising on dark tourism consumption; and (2) it unravels how dark tourism experiences can afford transformational benefits by showing how visitors reflect and create meanings as a result of their emotional engagement (i.e. the various types and intensity of emotions triggered and felt) while at the dark sites.

The remainder of the paper is structured as follows. First, a literature review explains how dark tourism research has evolved from studying tourists' motivations to tourism experiences in order to better understand the phenomenon. By drawing on a multidisciplinary literature (i.e. psychology, tourism and service research), it also explains the concepts of experiences, emotions in experiences and how storytelling techniques can be used in experience design for eliciting emotions. Then, the methodology of the study is presented, followed by a critical analysis of the findings. The paper concludes by proposing a new adapted framework and discussing its implications for future research.

Literature review

Dark tourism: from a motivation to an experience focus

Dark tourism has been traditionally perceived as a niche form of tourism relating to tourists interested to visit places of death or associated with death (Lennon & Foley, 2000). Various typologies and frameworks of dark tourism have been developed based on a demand motivation approach recognising that interest in death might not be the sole or at all the motivation and reason for someone visiting dark sites (Foley & Lennon, 1996; Hyde & Harman, 2011; Raine, 2013). Indeed, different degrees of 'darkness' are found in frameworks that reflect a wide spectrum of dark tourists and sites. As research understanding dark tourists' motivations also helps in managing dark sites (Poria et al., 2006), studies in this field have mushroomed. Stone's (2006) typology is the most popular framework conceptualising the dark tourism spectrum. This was later adapted by Raine (2013) who profiled consumers according to their dark tourism motivation. Dore (2006) undertook a study on motivations and found four types of dark tourism experiences: (a) Commemorative, (b) Knowledge seeking, (c) Satisfying curiosity, and (d) Cathartic. Empirical findings (Hall et al., 2010) from an investigation of the ninety fifth (95th) commemorative event in Gallipoli revealed that the visitors' motives are mainly connected with nationhood. Winter (2011) identified knowledge and interest in history followed by general interest as the major motivation of cemeteries' visitors. By seeing visits to cemeteries as a form of dark tourism, Raine (2013) also identified the core element of devotion in the dark tourism experience. Slade (2003) suggests that not all pilgrims have the same motives; thus, it is essential to explore the motivational factors, a claim which is also confirmed by Biran et al. (2011), who noted that different motivations can produce different benefits. The same authors found three factors motivating visits to heritage sites: (a) a consumer's desire to learn; (b) see it to believe it; and (c) a desire for an emotional experience. In a further study of heritage sites, Magee and Gilmore (2015) found elements in heritage service settings that facilitate a co-creation process of an emotional connection between the consumers and the sites. In conceptualising a framework to capture consumer experiences, Magee and Gilmore (2015) also identified three primary motivations, classifying tourists as: Knowledge seekers; Identity reinforcers; and Curious.

Overall, research in dark tourists' motivations reveals a multiplicity and diversity of motivations that go beyond interest in death. This in turn has raised several questions in terms of the differentiation and distinctive nature of dark tourism in relation to other forms of tourism such as, heritage and adventure tourism (Light, 2017). Findings also imply that dark tourism is more a form of an experience rather than a motivation, while it is the consequences rather than the drivers of this experience that matters more (Stone, 2012). Consequently, diverting from earlier research in dark tourism, the recent literature highlights the need to adopt an experience focus for better understanding dark tourism (Zheng et al., 2019; Light, 2017; Stone, 2012). The mortality mediation model by Stone (2012) has been proposed as a new way for theorising and understanding dark tourism consumption, because it focuses not on the motives for visiting dark sites but on the subjective experiences of visiting such places and the visitors' impacts of these experiences. The mortality mediation theory suggests that the dark sites provide and communicate moral meanings, which enable the visitors to engage with, reflect and negotiate moral concerns; however, the way and the degree to which the visitors may engage vary depending on their cultural and societal backgrounds as well as the context of their visit. Thus, the mortality mediation theory provides a new psycho-analysis approach for conceptualising dark tourism from an experience focused perspective, as it views the latter as a personal and individual experience rather than an inherent characteristic of the site or place. This is also a more flexible theory going away from the assumptions of past dark tourism research that uncritically impose the 'western way of thinking' of dark tourism in various cultural and geographical contexts (Light, 2017). Instead, the mortality mediation theory allows and requires an understanding of the various and constantly changing meanings and interrelations of death, dying and the dead in various contexts and times (Stone, 2012).

The mortality mediation theory represents an experience focused approach for conceptualising dark tourism that is also in line with current service research investigating the nature and the concept of customer experience. For example, the service research literature advocates that experiences are social practices that are phenomenologically, subjectively and contextually co-created by the involved actors (Jain et al., 2017; Smit & Melissen, 2018; Kranzbühler et al., 2018). The customer experience is often considered as a subjective and internal response in the mind of a person who is engaged on an emotional, physical, intellectual, and even spiritual level (Dou et al., 2019). The customer experience is also considered as both the process (including the ongoing customers' perceptions, emotions and direct observations) and the outcome (referring to the accumulating knowledge, skills, emotions, and attitudes) (Jain et al., 2017) of a customer engagement with the service ecosystem. The customer experience is also perceived as a formative structure (representing the customer interaction with an environmental context), and a reflective structure, reflecting the customer reactions to the environmental context elements and actors (Kranzbühler et al., 2018). Indeed, recent studies (e.g. Chen & Dean, 2020) reinforce and highlight the critical role of the experience context in enabling and/or constraining the actors' engagement in co-creating and appropriating the impacts of their experiences. Overall, the customer experience is regarded as a set of emotions, perceptions, and attitudes, which are shaped as a result of interacting with the actors, the processes and the environment of the service context/ecosystem and which lead to multi-dimensional customer engagement including the behavioural, sensory, emotional, cognitive, and/or spiritual reactions of the actors (Jain et al., 2017). Consequently, customer experiences can be conceptualised in two ways: (1) in a static way examining the cognitive, emotional, and sensory customer engagement captured at one or more interaction touch points during the experience; and (2) in a dynamic way, that considers the cognitive, emotional, and sensory customer engagement during and over the whole duration of the customer journey, i.e. even well beyond the service consumption.

In a similar vein, research in dark tourism has also recognised the multi-dimensionality of dark tourism experiences. For example, MacCarthy (2017) identified the various facets of dark tourists' engagement including physical, sensory, restorative, introspective, transformative, hedonic, emotional, relational, spiritual and cognitive reactions. Current literature also recognises the transformative, subjective and phenomenological nature of dark tourism experiences. Zheng et al. (2019) found that many more visitors now get more deeply engaged, as they see their visits as opportunities for (re)-connection, understanding, meaning-making, self-exploration, self-developmental, transformative and/or even life-changing experiences. Given these variations in consumer motives and emotional behaviours, it is important to continue and expand research on consumer motivation by specifically looking at how the emotional connections of visitors are formed and they influence the consumers' engagement and experience at the dark site (Cheer et al., 2017). Dark tourism research also emphasises the subjective and phenomenological nature of dark tourism experiences. For example, many dark tourists do not also accept sanitised presentations of history and they experience dark places in their own way by accepting, enriching or rejecting messages and stories (Smith, 2012). Peoples' own interpretations, reflections, understanding and retrospective experiences of dark tourism enable them to make their own meaning of dark places and death, which sometimes may be different from those intended from tourism supply (Du et al., 2013).

Dark tourism research also offers some preliminary findings about the static and the dynamic nature of dark tourism experiences, as it distinguishes between: the sensorial, emotional and cognitive reactions of visitors at the moment of the visit; and the dynamic reflective, learning, transformational and long-term impacts that dark tourism experiences can generate. For example, of particular importance are the emotions elicited by dark tourism experiences and their affordances to trigger reflection, transformative and spiritual impacts for dark tourists (Light, 2017). This is mainly because of the special nature and the environmental elements of the dark tourism (sites/context) that can trigger visitor engagement in various dimensions; e.g. dark sites are special places that evoke intense emotional experiences (Hede & Hall, 2012) and they frequently represented sacred places with spiritual meanings (Hall et al., 2018). The mortality mediation model also highlights the transformational and self-developmental opportunities afforded by dark tourism, as the particular emotional engagement elicited by dark tourism experiences, can engage the visitors into a reflection and contemplation of the nature of mortality (Light, 2017). For example, by studying the mixed emotions felt by visitors at one of the darkest tourism places – the Memorial Hall of the Victims in Nanjing Massacre – Zheng et al. (2019) showed how the negative ambivalent emotional experiences elicited by the visitors trigger them to engage in a learning process that helped them convert this emotional experience into a meaning-making process. In fact, findings showed that it was the negative rather than the positive emotions driving a negotiated and transformed meaning making process. Moreover, although emotions are not the unique or defining feature of dark tourism, the nature of these emotions has been proposed as a way to differentiate dark tourism from other forms of tourism (Nawijn et al., 2016). In this vein, research investigating the dark tourists' emotions can significantly contribute to our understanding and research about dark tourism conceptualisation.

Affect is another concept related to the emotions that has also emerged within tourist studies. Although often used interchangeably with emotion, affect refers to the imperceptible, visceral and embodied ways in which people are affected by place before their conscious awareness of it and before they form an emotional response (Golańska, 2015). Although places associated with death afford to elicit and produce affective responses amongst visitors, only one study has examined, so far, the affective dimensions of dark tourism experiences (Buda, 2015b). Light (2017) identified the affective dimensions of dark tourism experiences as an urgent research priority for understanding dark tourism and the way in which people consume dark tourism. This study will use the terms affect and emotions interchangeably by highlighting the emotional response and the reflection that takes place when contemplating the nature of mortality (Stone, 2012).

Although research has explored the role of emotions in dark tourism experiences (Lagos et al., 2015; Dore, 2006; Hede & Hall, 2012), there are still various literature gaps and academics call for more research aiming to better understand the visitors' emotional experiences (Straker & Wrigley, 2016; Prayag et al., 2017). Researchers (Light, 2017; Zheng et al., 2019) also advocate the study of the emotions and affects generated by dark tourism as a critical research priority, because of the affordance of the former to trigger introspective, sensory, transformative and spiritual dimensions of the experiences, whose exploration and understanding are also still lacking. Research in how meanings are generated from tourism experiences is still in its infancy (Packer & Gill, 2017). Tourism research is also lacking investigating the factors (such as, story-telling, art-based interventions, and cultural landscapes) that can elicit personal meanings and transformation (Sigala, 2019). Furthermore, the mortality mediation theory (Stone, 2012) and its role in generating meaningful tourism experiences at dark tourist sites still awaits fuller investigation and development (Zheng et al., 2019).

This study aims to respond to these calls for a better understanding of the complex dark tourism phenomenon (Hall et al., 2018; Hyde & Harman, 2011) by adopting an experience-focused perspective that aims to unravel the role of the visitors' emotions and emotional engagement while at the dark site on their dark tourism experience. The study also considers the role of story-telling techniques (used by many dark sites for communicating and presenting messages) in eliciting the visitors' emotions, interactions, cognitive learning and reflective processes that may in turn provide them with transformative impacts.

Emotional engagement

The role of emotions in consumption experiences is widely discussed in marketing (Bagozzi et al., 1999), psychology (Bader, 2016) and tourism (Servidio & Ruffolo, 2016). Various tourism studies have focused on affective tourism (Buda, 2015a, 2015b); the link between emotions and destinations

(Prayag et al., 2017) and recently on emotions and heritage sites (Oren et al., 2021). Research in specialty or niche tourism (Novelli, 2005) has explored various issues related to emotions including: emotional responses to yoga (Dillette et al., 2019); customer experiences using digital channels (Straker & Wrigley, 2018); the use of emotional language in dark tourism online promotion (Lagos et al., 2015); emotions as digital drivers of consumer experiences (Wrigley & Straker, 2019); emotional reactions to adventure tourism (Faullant et al., 2011); secular pilgrimage experiences (Hall et al., 2018); and walking tourism experiences (Kato & Progano, 2017). Some researchers distinguish emotional dimensions as positive or negative (e.g. Prayag et al., 2017), but most studies focused on the positive rather than the negative emotions, while they have also been framed under the mindset that negative emotions have a negative impact on the visitors' experiences (Oren et al., 2021). Others (Servidio & Ruffolo, 2016) refer to emotions in a broader sense and the general relationship between the term and the tourist in particular settings. Overall, two issues are noteworthy and deserve future research: (1) how positive emotions may affect consumers differently when compared to negative emotions (Norman, 2004); and (2) research challenging the general mantra of negative emotions having a negative impact on experiences, while also demystifying the positive role of negative emotions on customer experiences by unravelling the mechanisms that trigger visitors to co-create value and transformational benefits.

Dark tourism research has confirmed the importance of emotional engagement in dark tourism sites (Frew, 2018; Hede & Hall, 2012; Dore, 2006), because emotions affect the humans' daily lives, their activities and subsequently their decision-making (Straker & Wrigley, 2015). In dark tourism, when consumers assign meaning to their own experience, this can have a high emotional impact (Cameron & Gatewood, 2003). Similarly, authors claim that when emotions are released during consumer engagement, individuals experience spiritual effects (Hall et al., 2018). This emotional engagement, where deep engagement is achieved during consumption, to the point where the visitor feels lost in time, has links with positive psychology (Seligman & Csikszentmihalyi, 2014).

Shaver et al.'s (1987) categorisation of the six basic emotions (love, joy, surprise, anger, sadness and fear) is the most popular and agreed emotional scale in the literature. In tourism research, scholars have also used this classification for categorising emotions; for instance, Prayag et al. (2017) focused on the positive emotions (surprise, joy, love) to explore levels of destination satisfaction and intention to revisit, while Straker and Wrigley (2018) explored emotions used in airport mission statements and found 'pride' to be the mostly elicited and represented emotive concept. Recently, Oren et al. (2021) focused on negative emotions and provided evidence of their positive impact on visitors' dark experience. Yet, their study has failed to unravel and explain the ways in which negative emotions can trigger and support value co-creation in a (dark) experience setting.

In this study, the term *emotions* refers to the *subjective feelings* and their *intensity* (Bagozzi et al., 1999) felt by the consumer when experiencing a phenomenon. Shaver et al.'s (1987) basic categories of emotions will form the basis for exploring the relationship between subjective feelings and dark tourism experiences. Special attention is paid on the negative emotions elicited to visitors at dark sites and on the mechanisms and the ways in which these negative emotions trigger and foster the visitors to create and perceive transformational value, benefits and satisfaction from visits at dark sites.

Linking emotions and storytelling

Research in emotions suggests there are various factors that can trigger either positive or negative emotions (Ding & Tseng, 2015). One such factor is storytelling. Stories have long been discussed as influencing consumers (Mossberg, 2008) and their emotions (Akinbode, 2013). In a story there is a plot, a setting and a narrator with the aim to create a final point to be made. Stories can become even more powerful when narrated in authentic settings by tour guides or knowledge experts (Cheal & Griffin, 2013). For example, participating at a heritage place such as Auschwitz, whereby experts narrate stories from the past can enhance emotional intensity due to its originality. Tourists are able to visualise the story by standing at the original place where the building is found. Therefore, the role of tour guides accompanying group tours during the emotional experience cannot be overlooked. Stories can also be created and curated through original photos or videos (Rose, 2016). For example, in Washington DC the Titanic story was revised based on original photos, authentic items and stories showcased next to those items. The museum in Gettysburg narrates the battle of Gettysburg in a separate room where the 360 degree painting of the battle is showcased. Such heritage experiences are also able to produce different emotions (Hede & Hall, 2012) and meanings to visitors (Packer & Gill, 2017). This paper argues that the visitors' emotional intensity may be enhanced when original storytelling is involved.

As outlined above, personal motivations for visiting heritage sites may result in various visitors obtaining different benefits (Poria et al., 2006); hence, different types of emotions may also be elicited. For example, heritage tourists who have personal connections with Auschwitz may feel extreme sadness when listening to knowledge experts narrating original stories onsite. Alternatively, a leisure tourist who is driven to the same site by curiosity, may feel shock at unravelling the story in front of the authentic site. Visits to heritage sites may also reinforce identity (Timoney, 2019). Therefore, it is important to clarify the type of emotional connection consumers feel during such experiences. Building on the work of Poria et al. (2006), who contend that different motivations may lead to different benefits, it is important to explore the emotional connections felt by different consumers, especially when stories are narrated during the experience, and people may interpret and (emotionally) react and engage to stories differently. The study attempts to further explore the influence of storytelling on the visitors' emotional engagement with the dark tourism site. It is anticipated that the emotional engagement experienced and recorded would contribute to Stone's (2006) original typology of dark sites.

Research methodology

Research aims

The study aimed to provide a deeper understanding of dark tourism by adopting an experience-focused perspective in order to investigate the visitors' emotional and transformative experiences elicited by their consumption of dark places. The study adopted a deep reflection approach [based on Van Manen's (2016) phenomenological practice] for understanding the type of the emotions elicited through the story-telling techniques and communications provided at three dark tourism sites. As this is an exploratory study, the concepts of emotion and feeling are used interchangeably.

Research approach

This paper is an autoethnographic exploration of the reflective experience of one researcher (Khoo-Lattimore, 2018) of three dark sites. Magee and Gilmore (2015) claim that heritage sites are suitable places for reflection. Autoethnography is a narrative research approach that describes and systematically analyses (graphy) personal experience (auto) in order to understand cultural experience (ethno) (Ellis et al., 2011). It is a 'context-conscious' approach, as the researcher is both the 'subject' performing the investigation and the 'object' of the investigation (Ngunjiri et al., 2010).

According to Adams et al. (2015, p. 45)

> The term autoethnography invokes the self (auto), culture (ethno), and writing (graphy). When we do autoethnography, we study and write culture from the perspective of the self. (...) we look inward-into our identities, thoughts, feelings and experiences-and outward-into our relationships, communities, and cultures.

Thus, autoethnography is a qualitative method that uses the researcher as the subject of exploration while displays narratives of his/her personal experience (Ellis & Bochner, 2000). Consequently, the personal experience invites the researcher to 'think with the evocative stories you read, reacting and reflecting with all your senses' (Bochner & Ellis, 2016, p. 9). Its major strength is that it is a powerful analytical process that can create evocative narratives that illuminate socio-cultural processes and understandings from a deeply personal perspective by using a back and forth gaze to focus outward on those socio-cultural contexts while also looking inward at the vulnerable self.

Autoethnography is heavily adopted in emotional intensive contexts (e.g. illness death and loss), as it provides for engaging closely with emotion-laden and highly sensitive topics (Chang & Horrocks, 2006), which is also the case for dark tourism research. Autoethnography is also an appropriate approach for examining the role of emotions on the conceptualisation of dark tourism, since previous research has shown the role of emotions in defining and determining the nature and the impact of visitors' emotions on their dark tourism experiences.

Overall, autoethnography is a systematic and intentional pursuit of an understanding of the self within a socio-cultural context, and recording interpretations as the experience unfolds. To achieve that, the researcher followed Van Manen's (2016, p. 17) phenomenological practice ['gravitates to meaning and reflectivity'] in order to remain reflective throughout the experience and to bring meaning to life. This is because it is not the content of the narrative/story itself that contributes to understanding in autoethnography, but the reflective processes elicited by capturing, telling and analysing these narratives in a similar way to which psychiatrists urge patients to self-reflect and self-diagnose their emotional issues (Bochner, 2001). In this vein, the researcher ventured on a journey of phenomenological practice (Vagle, 2018) where reflection was prominent during the visit (Magee & Gilmore, 2015). The researcher also examined herself as the data and gave voice to her own emotional experience (Ellis, 1993). Hence, the autoethnography of this study focuses on the essence of the experience and captures individual meanings, which may vary among individuals (Vagle, 2018). To that end, the researcher (providing the autoethnographic account) took a phenomenological approach to dark sites by focusing on the meanings deriving from her reflections, while also remaining open to her emotions elicited by her personal engagement with the dark tourism sites.

The phenomenological approach of autoethnographic research is appropriate for this study, as dark tourism experiences are argued to be subjectively, phenomenologically and contextually determined.

There are mixed arguments about autoethnography as a research approach. Delamont (2007) summarised six arguments against autoethnography:

> it cannot fight familiarity, it cannot be published ethically, it is experiential and not analytic, it focuses on the wrong side of the power divide, [...] it abrogates our duty to go out and collect data, [...] [and] we are not interesting enough to be the subject matter of sociology.

In their meta-analysis study, Olmos-López and Tusting (2020) showed why the critic by Delamont (2007) is not valid, as it dismisses the work of introspection, the value of 'making the familiar strange', and the process of analysing one's own experiences in a social and cultural context in a way that can provide understanding of issues others might be facing.

Part of the criticism about autoethnography also relates to narratives being too thin (Marcus, 1998). To address this and ensure that the capture and analysis of narratives are content rich and deep, the researcher took notes and photos to support her later reflections, while throughout the experience she engaged in an intensive self-critical process (Mkono, 2016) and internal dialogue (Sadala & Adorno, 2002; Dahlberg et al., 2008). In addition, the narrative of the personal experiences was conducted immediately after her return while memories were still fresh, while their analyses were supported by consulting the notes taken at the field and by recalling on the felt emotions (Ellis et al., 2011). The researcher's motivation to record her journey in a detailed and multimedia format as well as her high engagement with dark sites were driven by her prior interest in dark tourism research and a sense of responsibility to engage in ongoing reflexive practice (Khoo-Lattimore, 2018). Accordingly, she was practically and mentally prepared to immerse herself in the experience (St-James et al., 2018) and to use an autoethnographic approach for data interpretation.

It is worth noting that all the visits took place over one week. Although this was done for logistical reasons, it also allowed the researcher to fully immerse and engage herself as an active participant in the experience (Pine & Gilmore, 1998). Due to the short timeframes between the visits, the researcher engaged solely on dark sites, which allowed feelings to intensify (Prayag et al., 2013). and excluded herself from any other leisure and social type of experiences, which may blur or perplex the generation and impacts of feelings.

Data analysis

Autoethnography is a qualitative method that allows the researcher to interpret his/her own experience by reflection (Spry, 2001). During the reflective experience, the researcher made a conscious attempt to remain open to the phenomenon (Van Manen, 2016). The reflection is strongly tied with individual meanings and adopts a constant process of meaning-making exchange between the researcher and the lived experience (Dahlberg et al., 2008). The individual reflections included notes and photos and more notes following post-experience reflection. Reflections and meanings were drawn from photos (Packer & Gill, 2017; Sidaway, 2002), story elements and sounds (Sigala, 2019), that formed part of the experience. Journal writings were read numerous times during the lived experience and the researcher took time to reflect pay attention to her emotions as these are important during the reflection process (Van Manen, 2016); Being an active participant meant the researcher engaged her mind and became fully aware of the moment of 'now' (Akinbode, 2013). The journal writings focused on the impact that visitor elements had on her while she was living the experience. The researcher engagement with the dark tourism phenomenon will also highlight practical implications in terms of designing emotional experiences.

Research context

The researcher collected data and reflected on her experiences by visiting three different dark sites. Reflecting on dark tourism experiences from three different dark sites is critical, as previous research shows that the dark tourism experiences and the visitors' engagement with the dark site depend on the context of dark tourism. The dark sites of the study include:

- Ground Zero is the place where the tragedy and the heroism unfolded when the planes hit the twin towers on September 11, 2001. The memorial that stands there today is a symbol of the community's strength having undergone one of the biggest attacks in history (911 Ground Zero Tours, 2020).
- Gettysburg is a park where the most impactful civil war took place in 1863 with more than 50,000 casualties over three days (July 1–3). President Lincoln's famous address declared the struggle for freedom and democracy. Gettysburg today symbolises the greatest democratic battle in American history (Gettysburg Foundation, 2020).
- Ellis Island is an island outside New York which was used to process immigrants with approximately 12 million immigrants having gone through the island in search for a new life between 1892 and 1954. Now a museum since 1990, the island narrates the story of American immigration (The Statue of Liberty Ellis Island Foundation, 2020).

Presentation and discussion of the findings

Reflections on dark tourism experiences

The following sections reflect the researcher's level of emotional engagement in the three dark sites and will be narrated in first person; Ground Zero was rated as the darkest tourism experience based on the intensity of the researcher's emotions.

Ground Zero (New York, USA)

To me Ground Zero was the most intense emotional experience. This was due to a combination of activities that took place: the narration by a local tour guide in a small walking group, whose nickname was Betty; the opportunity I had to reflect on the symbolic meanings of objects and site; authentic elements, such as storytelling by a local who had lived the events; objects and sites which maintained their originality to this day in the museum. Betty's narration took me back to the morning of 9/11 where she was, in a hotel when she heard about the explosion. Immediately, I felt emotionally engaged and taken back with her to that hotel. The story continues with people gathering close to each other waiting to hear news of the events that took place. I was really focused on Betty's words and, despite knowing the events that took place via the media, I felt I needed to hear it from a local who lived the experience. Betty's storytelling was clear and at a pace that allowed me to reflect and relive the story. I started reflecting on meanings for the local people as the group started walking to symbolic sites. I noticed there was silence among us and it seemed everyone was reflecting. Betty pointed at the key monument, Ground Zero, assigning her own interpretations of what it means for the local community: the hole in the middle can never be filled due to the pain it has caused. Emotions of anger, sadness and empathy were released to the American people suffered the loss of loved ones. A collective grief that will never heal. 3000 innocent lives suddenly got lost in one day is all I could reflect upon the interaction with the site. I stood still allowing my anger to surface and tears flow from sadness. These mixed emotions intensified as Betty continued with her genuine storytelling techniques. Allowing the group to reflect, I took the opportunity and felt lost in that moment and total grief for the loss of life. Then Betty pointed to the tree that had survived the attack and narrated it was a symbol of hope. A sudden feeling of awe and disbelief at what I could hear. A tree that stayed alive? I suddenly felt a sense of connection with the place and Betty as I understood what it meant for her. A sign of hope for the future. To me, the tree was a sign of welcome into a place that will always show love to people. It was a real connection. Positive emotions started to arise: awe, admiration and love for the community that is protecting the root system of this tree with highly protected cameras, so that it survives. A highly yet small emotive item was a live rose emerging from someone's name engraved on a plaque; that fateful day was the person's birthday. The intention of the community is to ensure that the birthdays of those who died on September 11 will never be forgotten. I felt a deep emotional engagement at that moment picturing families visiting their loved ones every year to commemorate them and show respect. I focused on the rose and the name on the plaque and suddenly felt a deep connection with that person, who had lost their life in tragedy. I felt another tear flow and promised I will never forget and will always remember them.

Gettysburg battlefield tour (Pennsylvania, USA)

I chose to visit Gettysburg as it formed part of my reflective study on heritage tourism, which was part of a larger study. The bus tour to Gettysburg comprised a ninety-minute drive from Washington DC. The driver had the role of tour guide and narrated the story during the trip. Known to us only as 'Roger', the guide had a wealth of knowledge. The Battle of Gettysburg in 1863 lasted three days (July 1-July 3), with a loss of over half a million lives. I stayed focused on the part of the storytelling, which included a particular family, the Shriver family who played a crucial part in Gettysburg. As the driver approached the historic town of Gettysburg, Roger surprised us with a stop outside the original house of the Shriver family. I felt excited to see the original Shriver home and most importantly that we were allowed to go in even though it was not on the itinerary. My excitement was shared among the group as we had all shared similar emotions of awe about this particular family. I was amazed as I could clearly see the original holes made by the bullets that hit the house. I suddenly felt awestruck by the originality of the property and the tour guide that was waiting for the group dressed in clothes of the era calling herself Miss Margaret and preparing the group for storytelling. A smile showed at my face as I felt I will be brought into the story. Miss Margaret took us through each room of the house and explained how the family lived before the war broke out, and what happened afterwards. The rooms included original artefacts from the soldiers including symbolic elements such as clothes with bloodstains, guns, messy tables and broken glasses. I admired the effort by the story creators to maintain authenticity in the original household as it portrayed elements that reflected the everyday life of a family of the era. Miss Margaret was speaking in an original tone of that era and this enhanced my positive feelings throughout her story-telling efforts. My curiosity intensified as I wanted to hear how the story ended but Miss Margaret smiled at me suggesting it will be revealed soon, when we reach the last room. Mr Shriver never returned to see his wife, children and a new born child, and that made me feel sad as I reflected on his poor wife trying to raise a family on her own in those times. To me it felt it was a story of the only civilian who became a soldier feeling a sense of responsibility to protect his family and community, yet never returned. It made me reflect on how life can easily change forever beyond one's control. This progressive storytelling created emotional intensity varying from mixed emotions of shock about the number of deaths in three (3) days of war, sadness and love about a tragic family story of family of the nineteenth century who underwent the unfortunate circumstances of a civil war.

Ellis Island (New York, USA)

Ellis Island is just a short ferry trip from New York. It is not featured as a popular site in tourist brochures. This may be due to the fact that there have not been as many deaths on Ellis Island, and deaths were caused mainly by disease. The building is maintained in its original state, with furniture placed in exact spots during the key years of immigration, 1909–1911. I was personally motivated to visit a site with a story of immigration as I was also a second-generation migrant in Australia. When I walked into the heritage building I felt a sense of history which took me back to the years of immigration. Whilst there was no tour guide to recreate the personal stories, the rooms were filled with images and sounds that narrated personal stories of migrants. My journey began on ground floor and then gradually moved to level 1 and 2. As I moved into different rooms and displays, I focused on photos picturing families undergoing medical tests and examinations and short stories narrating what would happen if they were sick. Families with sick members were pulled apart and some women never saw their husbands again. I felt sad reading such stories of people who underwent these hard conditions in addition to undergoing discrimination. As a second generation Greek being an Australian migrant I reflected on how lucky I was to be successful in migrating in the current times. Moving from room to room brought me into the setting where stories became alive with sound reincarnations. I felt admiration and the successful display of the immigration history of USA maintaining story authenticity. I was struck with awe by a prominent statue of the first female who settled in America and felt proud of how women fought so hard to influence change. I reflected on how lucky I have been being accepted in my community as equal. I felt I needed to show my respect to this woman, a symbol of female representation who fought for me to enjoy an equal life. Her image will stay in my mind forever. As a researcher Ellis Island stood as a symbol of remembrance of the dark journey for some migrants who underwent hardship into being equally accepted into their new communities.

Summary of autoethnographic reflections

The researcher was involved with phenomenological inquiry and this experience gave her a new lens through which to view the dark tourism experience by seeking the essence of the experience and assigning her own interpretive meanings. Although the researcher had had previous exposure to dark sites, these experiences in the U.S.A. were impactful. Ground Zero was found as the site that had the highest emotional engagement in terms of the type and intensity of the emotions being elicited and their resulting impact in generating reflections, critical thinking and transformation. This was also the site where she experienced strong empathy for the local community and created new perspectives on the threat of terrorism and the need for closer communities. Mixed emotions of anger, sadness and deep love were formed onsite (Frew, 2018) being fully immersed in senses and listening to the authentic storytelling (Sigala, 2019) by the tour guide. These feelings lasted throughout the guided tour and remained intensified following the site visit as the researcher recorded notes of deep love for the American people.

Gettysburg triggered emotions of sadness with the loss of life as a result of the Civil War but also admiration at the pride of the American people to keep their history alive and convey their own meanings to the world. As the Civil War happened over 150 years ago, the story was interpreted by locals as a historic event that had symbolic meanings of identity. The researcher felt the tour to Gettysburg became an immersive learning-based experience (Pine & Gilmore, 1998) which narrated symbolic meanings of unity and national identity. Engaging deeply with the stories in Gettysburg contributed less towards intensifying emotions; instead the feelings that were created were an appreciation of identity (Timoney, 2019); joy for the originality of the story narrated by a tour guide in a simulated context; and respect and appreciation of history. The researcher interpretations were a result of contemplating death at various dark places (Stone, 2012) and interacting with authentic elements (Magee & Gilmore, 2015). Being fully immersed in the experience (Pine & Gilmore, 1998) and assigning individual interpretations at different dark sites based on storytelling, authentic elements and emotional engagement (Straker & Wrigley, 2015) created impactful meanings and an overall respect for the American history and people.

Finally, the researcher developed empathy (Frew, 2018) for those who had spent time on Ellis Island. As a second-generation Australian migrant, born to Greek parents, she heard stories of relatives who had to stay in a similar facility, Bonegilla, in north-eastern Victoria before being allowed to settle in Melbourne (capital of Victoria, Australia). The researcher felt an identity connection (Timoney, 2019) and an appreciation for preserving an important part of history for migrants; while she acknowledged the cultural importance and identity links for the U.S.A. The researcher assigned her own reflections to current global issues of migration and the risks that people are willing to take when seeking a better life. Engaging deeply with history and the stories produced new personal meanings and an overall meaningful cultural experience (Packer & Gill, 2017).

The dark tourism experience elicited strong emotional engagement of the researcher with the dark sites, which in turn lead to the generation of various critical reflections, critical and transformative thinking of herself and the society. Before the experience, the researcher had feelings of anticipation and excitement as she sought a meaningful experience where she would learn and feel in an immersive context. The most intense emotional engagement was 911 Ground Zero. The individual interpretations of symbolic meanings based on authentic objects (such as the tree that had survived the disaster) increased emotional intensity to deeper levels. The sense of respect that was developed in Gettysburg was a result of appreciating the past heroes who fought for the human values of freedom. This knowledge-based experience enhanced learning and allowed the researcher to immerse herself in a context which combined authentic storytelling elements (Mossberg, 2008). The deep reflection felt on Ellis Island was interpreted as a result of her own history as a migrant. A sensory engagement which contributed towards feelings of surprise and awe at the originality of the site and the preservation of recent history. The researcher walked away feeling spiritual (Hall et al., 2018). This feeling was a result of her own engagement with the tour guides, the locals who have a passion for their own history and who show a desire to take visitors into their world, so they can live the experience (Van Manen, 2016). Dark tourist sites facilitate personal reflection (Magee & Gilmore, 2015).

Overall, the findings show that dark tourism experiences can elicit both positive and negative emotions before, during and after the experiences, which confirms prior studies on the role of emotions in travel (Prayag et al., 2017). Overall, the study findings enrich this literature by (Table 1): (1) identifying the dimensions of the visitor emotional engagement (i.e. the type and the intensity of emotions) at the dark tourism site; and (2) unravelling how these dimensions determine the type (i.e. darkness) of the visitor tourism experience.

Table 1. Levels of emotional engagement at dark tourism experiences.

Emotional engagement (type and intensity of emotions)	Ranking of the researchers' experiences at each dark tourism site based on the level of *emotional engagement*; 1. Deepest emotional engagement 2. Deep emotional engagement 3. Some level of emotional engagement
Deep emotional engagement Mostly deep and intense mixed positive and/or negative emotions (shock, anger, sadness, love, admiration) *Limited emotional engagement* Mostly positive feelings	1. Ground Zero 2. Ellis Island 3. Gettysburg

The study also expands our knowledge by providing evidence of the positive relation between emotional engagement and the transformational impacts generated for the visitor.

Finally, the study contributes to the field by illustrating the role of the following factors in eliciting the visitors' emotional engagement: (1) story-telling; (2) interactions with authentic objects or characters; and (3) the visitor's background and previous lived experiences allowing him/her to make self-connections with the (stories of) dark-site and/or to transfer the symbolic meanings of the (story of) dark site to other personal or contemporary issues. Overall, the findings identify and validate the emotional engagement and the personal interactions and reflections (triggered through story-telling and the symbolic meaning of objects) as the mediating processes leading visitors to co-create their subjectively formed and understood dark tourism experience (Table 2).

A revised framework for understanding dark tourism (experiences): its theoretical and practical implications

Based on the findings of the personal reflections and narratives, the study proposes a new framework for conceptualising dark tourism. The new framework recognises the role of the visitors' emotions in determining their type of dark tourism experiences and it is based on two emotional dimensions (Figure 1); the type of the emotions elicited by the experience (Straker & Wrigley, 2015); and the intensity/grading of the emotion when compared to Stone's (2006) dark shade framework. The framework builds upon and further expands Stone's (2006) and Raine's (2013) model, which categorised dark tourism experiences based on the type of tourists visiting those sites and

Table 2. Factors eliciting the visitor emotional engagement during the dark tourism experience.

Factors eliciting visitors' emotional engagement	High level	Low level	Ranking of the researchers' experiences eliciting emotional engagement based on the Stone's (2006) shades of dark tourism: 1. Darkest 2. Darkish 3. Dark/Light
Interactivity with authentic site	High level of personal interactivity with authentic site	Low level of personal interactivity with authentic site	1. Ground Zero 2. Gettysburg 3. Ellis Island
Interactivity with story-telling	High level of personal interactivity with story-telling	Low level of personal interactivity with story telling	1. Gettysburg 2. Ellis Island 3. Ground Zero
Reflection on symbolic meanings	Deep reflection on symbolic meanings Intense contemplation on death, living and life	Limited reflection on symbolic meanings Limited contemplation on death, living and life	1. Ground Zero experience 2. Ellis Island 3. Gettysburg

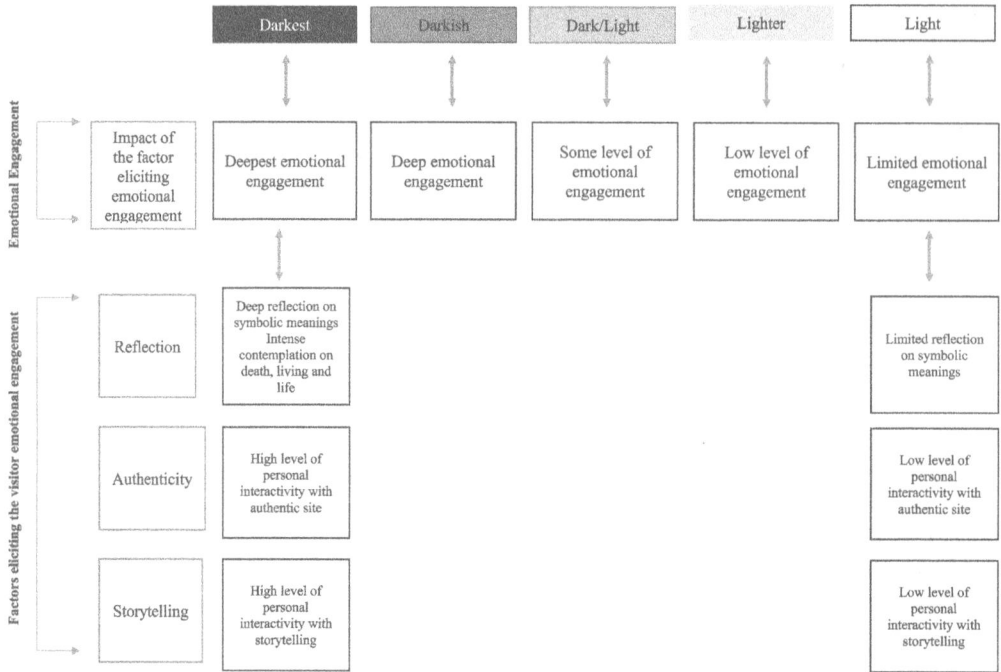

Figure 1. Conceptualising dark tourism: An extended dark tourism experience spectrum based on the visitors' emotional engagement elicited by three factors used at dark tourism sites (adapted from Stone, 2006; Raine, 2013).

their motivational factors. This study advocated and provided evidence of the role of the visitors' emotions in influencing their dark tourism experience both in terms of the nature of their emotional engagement (i.e. type and intensity of emotions) as well as of the factors eliciting the latter (e.g. story-telling, authenticity, reflection on personal experiences and background). In this vein, this study and the proposed framework are also in line and confirm the subjective and contextual nature of experiences (Chen & Dean, 2020), as the latter are co-created based on the actors' interactions and engagement with the service environment and their subjective interpretations and meanings of the latter. Thus, the study contributes to the literature by providing a framework that better explains and categorises dark tourism (experiences) based on the level of the emotional engagement that the visitors elicit and subjectively understand by different types of dark tourism experience environments.

Overall, the proposed framework identifies different levels of emotional engagement and forms of reflection that relate to different dark shades of experiences, which in turn corroborate past studies on the various forms of nuanced dark tourism experiences. The framework also identifies and provides evidence of the factors (storytelling, site interactions, reflections and symbolic meanings of objects) eliciting the visitors' emotional engagement and enabling them to create meaning in dark tourism experiences (Seaton, 2002; Lennon & Foley, 2000). The proposed framework (Figure 1) also refers to the study findings by showing how the researcher's experience at the three dark tourism sites provides evidence of the interplay between the level of visitors' emotional engagement, the factors eliciting emotional engagement and the type of 'darkness' of the experience.

The framework advances research in dark tourism, because: it adopts an experienced-based approach for better understanding dark tourism, which approach was missing from previous research (Light, 2017); it identifies and provide evidence of the role of emotions in influencing dark tourism experiences and their ability to generate (transformational) impacts for visitors (e.g. Zheng et al., 2019); and it adopts a multidisciplinary approach highlighting the subjective and contextual nature of experiences for conceptualising dark tourism (experiences) based on the phenomenological and socially constructed way in which people contemplate death, living and life when interacting with dark tourism contexts (Stone, 2012). The latter is particularly important as previous research has been accused of using and transferring the 'western' conceptualisation of dark tourism in other cultural (Asian) contexts without reflecting and questioning whether and how the connections of living and the dead, the meaning of dying and death may differ across contexts (Light, 2017).

Consequently, the proposed framework has several theoretical implications and provides directions for better conducting cross-cultural research in dark tourism. This is because the framework identifies the factors and the subjective emotional and cognitive processes influencing the local ways in which people understand and reflect on death. By being more versatile and adaptable the new framework enables the researchers to more deeply explore the formation, the meanings and the dynamic changes of dark tourism experiences across a broader range of geographical and cultural contexts. From a practical perspective, the framework provides useful implications to professionals by: (1) identifying the factors (e.g. story-telling, authenticity of objects, creation of interaction touchpoints and technologies) that they can design and manipulate for delivering and eliciting different types of dark tourism experiences to various market segments; and (2) identifying the emotional and the cognitive processes that dark tourists use to interpret and create meanings from their experiences, so that the managers can design a much more responsible dark tourism site that 'caters' to the diversity and the 'sensitivities' of the various cultural backgrounds of its visitors.

Conclusions, limitations and ideas for future research

Although consumers have engaged with dark sites over the years, there is still disagreement in how these experiences should be defined (Hyde & Harman, 2011). The researcher's self-inquiry experience led her to agree with authors (Hall et al., 2018; Magee & Gilmore, 2015) that consumers may have life changing moments at heritage experiences especially where authenticity and storytelling (Sigala, 2019) are involved. To that end, this study adopted an experience-focused perspective and an autoethnography approach for collecting and interpreting personal experiences at dark sites for providing a better and fresh approach in understanding and conceptualising dark tourism. Based on the findings, the study proposed a new framework that addresses the previous gaps, while it further expands the literature.

The new framework allows for mixed emotions as these vary for individuals at various dark tourism sites. Accordingly, this study agrees with findings by Oren et al. (2021) that research in (dark) tourism experiences should also focus on negative emotions, while mixed and/or qualitative methods should be used for unravelling the mechanisms and the contextual settings under which negative emotions can have a positive impact on visitors' experience and satisfaction. This study contributes and expands this recent literature by studying the impact of negative emotions and providing evidence that it is the deep reflection and the intensity of the feelings that are triggered by the mixed and negative emotions felt by the visitors at dark sites, and which in turn enable them to derive individual meanings and value from their dark tourism experience. Dark site managers need to be aware of the factors and the processes that can trigger different visitors' emotions (Straker & Wrigley, 2015), so that they can better 'stage' and design their servicescapes and experiences in order to elicit visitors with 'appropriate' emotions at the right time and place. Dark site managers should also accept that the triggered emotions and the meanings drawn from the 'staged' experiences may vary depending on the background of every individual, and so, they might need to customise experiences and emotion triggers per customer group and / or educate their guides to be able to personalise their interpretations and the triggers to elicit visitors' emotions based on each individual. Future studies should explore the role of authenticity in the visitor emotional engagement at dark tourism sites. Accordingly, the visitor experiences at dark sites need to be themed with stories and authentic objects with symbolic meanings that can trigger deep reflection, generate intense emotions and allow for new meanings to be created.

The autoethnography adopted by this study also provides learnings and implications for other researchers. As a phenomenological inquirer, the researcher's journey to the three dark sites over a period of one week helped her to expand her personal views on humanity and to develop lasting positive feelings of empathy and respect. By engaging in a reflexive practice, the researcher also created a dialogue with herself (Van Manen, 2016) during contemplation of death (Stone, 2012) and questioned her own ideas and emotions. Practicing daily reflection and approaching issues with an open mind also help the researcher to develop his/her critical thinking (Khoo-Lattimore, 2018). As Van Manen (2016) claims, phenomenology is a sensitive practice where the researcher seeks meaning with an open mind.

As with every study, this research is not free from limitations. However, the limitations provide various suggestions and directions for future research. This is a qualitative study based on personal reflections and narratives of one researcher in three sites. This study is also based on findings from the emotional impact and the explored elements of personal transformation during the lived experience of one specific individual. Moreover, both the researcher (her personal experiences, emotions and interpretations) and the studied contexts (three dark sites in USA) are drawn from and represent western cultures, religions and mindsets. In this vein, future studies should also aim to refine, validate and/or expand the study's findings by using a greater and more varied sample of visitors and dark sites that can represent different or even conflicting cultures and backgrounds. For example, studies could investigate whether western dark sites elicit or not (negative or positive) emotions to Asian visitors and how the latter may reflect upon them and derive (transformational) meanings (and vice versa i.e. western visitors experiencing dark sites in Asia). Mixing the cultures of research subjects (visitors) and research context (dark sites) in order to study the role of emotions, reflections and meanings in dark tourism experiences is critically important, as this study has also confirmed the general knowledge that the previous lived experiences and the cultural values of the individuals play a major role in eliciting emotions as well as triggering and guiding reflections, because they are the lenses through which people see, interpret and respond to environmental stimuli.

Overall, the new proposed model, its dimensions and their constructs as well as their interrelations need to be refined, validated and tested by a greater and multi-cultural quantitative study in order to generalise results.

During the lived experience, authenticity (Magee & Gilmore, 2015) was also found to play a role in creating an emotional impact. In this vein, future studies can use the revised Dark Tourism Spectrum in order to explore and compare the role of interplay between the site's location, storytelling and authenticity between socially constructed contexts and personal journeys.

More studies are also required to ascertain the elements triggering the emotional intensity during the experience by identifying the ways in which both negative and positive emotions are elicited by various contexts. Moreover, as past research has shown that the social context and the presence of others can critically shape dark tourism experiences, future research should also investigate the role and the impact of the visitors' engagements and reflections not only with themselves, the dark site objects, their stories-messages and symbolic meanings, but also their interactions, dialogues and negotiations with other visitors.

Finally, the study has focused on capturing and analysing the researcher's reflections and emotional engagement solely during her visit. However, customer experiences can also be interpreted in a dynamic way (i.e. during the whole customer journey, before, during and after the experience) as well as the life changing impacts of experiences can also go beyond the experience consumption stage. Hence, future research should also adopt a longitudinal approach investigating the longer term and transformational impacts of emotional dark tourism experiences on the visitors' daily life practices, lifestyles, perceptions, behaviours and/or attitudes towards life. In dark tourism experiences, it is usually more important what comes after the experience, and so, research understanding and exploring the visitors' practices of remembering, reflecting, learning and behavioural change is of critical importance.

Disclosure statement

No potential conflict of interest was reported by the author(s).

ORCID

Marianna Sigala 🆔 http://orcid.org/0000-0001-8997-2292
Effie Steriopoulos 🆔 https://orcid.org/0000-0003-4615-5393

References

911 Ground Zero Tours. (2020). *Ground Zero and 911 memorial tours*. Retrieved August 30, 2020, from https://911groundzero.com/.

Adams, T., Holman Jones, S., & Ellis, C. (2015). *Autoethnography: Understanding qualitative research*. Oxford University Press.

Akinbode, A. (2013). Teaching as lived experience: The value of exploring the hidden and emotional side of teaching through reflective narratives. *Studying Teacher Education*, 9(1), 62–73. https://doi.org/10.1080/17425964.2013.771574

Ashworth, G. J., & Isaac, R. K. (2015). Have we illuminated the dark? Shifting perspectives on 'dark'tourism. *Tourism Recreation Research*, 40(3), 316–325. https://doi.org/10.1080/02508281.2015.1075726

Bader, O. (2016). Attending to emotions is sharing of emotions – a multidisciplinary perspective to social attention and emotional sharing. Comment on Zahavi and Rochat (2015). *Consciousness and Cognition*, 42, 382–395. https://doi.org/10.1016/j.concog.2016.04.012

Bagozzi, R. P., Gopinath, M., & Nyer, P. U. (1999). The role of emotions in marketing. *Journal of the Academy of Marketing Science*, 27(2), 184–206. https://doi.org/10.1177/0092070399272005

Biran, A., & Hyde, K. F. (2013). New perspectives on dark tourism. *International Journal of Culture, Tourism, and Hospitality Research*, 7(3), 191–198. https://doi.org/10.1108/IJCTHR-05-2013-0032

Biran, A., Poria, Y., & Oren, G. (2011). Sought experiences at (dark) heritage sites. *Annals of Tourism Research*, 38(3), 820–841. https://doi.org/10.1016/j.annals.2010.12.001

Bochner, A. P. (2001). Narrative's virtues. *Qualitative Inquiry*, 7(2), 131–157. https://doi.org/10.1177/107780040100700201

Bochner, A. P., & Ellis, C. (2016). *Evocative autoethnography: Writing lives and telling stories*. Routledge.

Buda, D. M. (2015a). The death drive in tourism studies. *Annals of Tourism Research*, 50, 39–51. https://doi.org/10.1016/j.annals.2014.10.008

Buda, D. M. (2015b). *Affective tourism: Dark routes in conflict*. Routledge.

Cameron, C. M., & Gatewood, J. B. (2003). Seeking numinous experiences in the unremembered past. *Ethnology*, 42(1), 55–71. https://doi.org/10.2307/3773809

Chang, K. H., & Horrocks, S. (2006). Lived experiences of family care-givers of mentally ill relatives. *Journal of Advanced Nursing*, 53(4), 435–443. https://doi.org/10.1111/j.1365-2648.2006.03732.x

Cheal, F., & Griffin, T. (2013). Pilgrims and patriots: Australian tourist experiences at Gallipoli. *International Journal of Culture, Tourism and Hospitality Research*, 7(3), 227–241. https://doi.org/10.1108/IJCTHR-05-2012-0040

Cheer, J. M., Belhassen, Y., & Kujawa, J. (2017). The search for spirituality in tourism: Toward a conceptual framework for spiritual tourism. *Tourism Management Perspectives*, *24*, 252–256. https://doi.org/10.1016/j.tmp.2017.07.018

Chen, T., & Dean, A. (2020). Co-creation and engagement: Maturing and broadening the horizon. *Journal of Service Theory and Practice*, *30*(1), 1–4. https://doi.org/10.1108/JSTP-01-2020-310

Dahlberg, K., Dahlberg, H., & Nystrom, M. (2008). *Reflective lifeworld research* (2nd ed.). Studentliteratur.

Delamont, S. (2007, September). *Arguments against auto-ethnography*. Paper presented at the British Educational Research Association Annual Conference, London, UK. https://www.leeds.ac.uk/educol/documents/168227.htm

Dillette, A. K., Douglas, A., & Andrzejewski, C. (2019). Yoga tourism – a catalyst for transformation? *Annals of Leisure Research*, *22*(1), 22–41. https://doi.org/10.1080/11745398.2018.1459195

Ding, C. G., & Tseng, T. H. (2015). On the relationships among brand experience, hedonic emotions, and brand equity. *European Journal of Marketing*, *49*(7/8), 994–1015. https://doi.org/10.1108/EJM-04-2013-0200

Dore, L. (2006). Gallipoli, a visitor profile. *Historic Environment*, *19*(2), 46–51.

Dou, X., Zhu, X., Zhang, J. Q., & Wang, J. (2019). Outcomes of entrepreneurship education in China: A customer experience management perspective. *Journal of Business Research*, *103*, 338–347. https://doi.org/10.1016/j.jbusres.2019.01.058

Du, W., Littlejohn, D., & Lennon, J. (2013). Place identity or place identities: The memorial to the victims of the Nanjing massacre, China. In L. White & E. Frew (Eds.), *Dark tourism and place identity: Managing and interpreting dark places* (pp. 46–59). Routledge.

Ellis, C. (1993). "There are survivors": Telling a story of sudden death. *The Sociological Quarterly*, *34*(4), 711–730. https://doi.org/10.1111/j.1533-8525.1993.tb00114.x

Ellis, C., Adams, T. E., & Bochner, A. P. (2011). Autoethnography: An overview. *Historical Social Research/ Historische Sozialforschung*, 273–290.

Ellis, C., & Bochner, A. P. (2000). Autoethnography, personal narrative, reflexivity. In N. K. Denzin & Y. S. Lincoln (Eds.), *Handbook of qualitative research* (2nd ed., pp. 733–768). Sage publications.

Faullant, R., Matzler, K., & Mooradian, T. A. (2011). Personality, basic emotions, and satisfaction: Primary emotions in the mountaineering experience. *Tourism Management*, *32*(6), 1423–1430. https://doi.org/10.1016/j.tourman.2011.01.004

Foley, M., & Lennon, J. J. (1996). JFK and dark tourism: A fascination with assassination. *International Journal of Heritage Studies*, *2*(4), 198–211. https://doi.org/10.1080/13527259608722175

Frew, E. (2018). Exhibiting death and disaster: Museological perspectives. In P. Stone, R. Hartmann, T. Seaton, R. Sharpley, & L. White (Eds.), *The Palgrave handbook of dark tourism studies* (pp. 693–706). Palgrave Macmillan Publishers Ltd.

Gettysburg Foundation. (2020). *Explore Gettysburg*. https://www.gettysburgfoundation.org/

Golańska, D. (2015). Affective spaces, sensuous engagements: In quest of a synaesthetic approach to 'dark memorials'. *International Journal of Heritage Studies*, *21*(8), 773–790. https://doi.org/10.1080/13527258.2015.1020960

Hall, J., Basarin, V. J., & Lockstone-Binney, L. (2010). An empirical analysis of attendance at a commemorative event: ANZAC day at Gallipoli. *International Journal of Hospitality Management*, *29*(2), 245–253. https://doi.org/10.1016/j.ijhm.2009.10.012

Hall, J., Basarin, V. J., Lockstone-Binney, L., Yusuf, A., Winter, C., & Valos, M. (2018). Spiritual values and motives of secular pilgrims. *International Journal of Consumer Studies*, *42*, 715–723. https://doi.org/10.1111/ijcs.12436

Hede, A.-M., & Hall, J. (2012). Evoked emotions: Textual analysis within the context of pilgrimage tourism to Gallipoli. *Advances in Culture, Tourism and Hospitality Research*, *6*, 45–60. https://doi.org/10.1108/S1871-3173(2012)0000006006

Hyde, K. F., & Harman, S. (2011). Motives for a secular pilgrimage to the Gallipoli battlefields. *Tourism Management*, *32*(6), 1343–1351. https://doi.org/10.1016/j.tourman.2011.01.008

Jain, R., Aagja, J., & Bagdare, S. (2017). Customer experience–a review and research agenda. *Journal of Service Theory and Practice*, *27*(3), 642–662. https://doi.org/10.1108/JSTP-03-2015-0064

Johnston, T. (2013). Mark Twain and the innocents abroad: Illuminating the tourist gaze on death. *International Journal of Culture, Tourism and Hospitality Research*, *7*(3), 199–213. https://doi.org/10.1108/IJCTHR-05-2012-0036

Kato, K., & Progano, R. N. (2017). Spiritual (walking) tourism as a foundation for sustainable destination development: Kumano-Kodo pilgrimage, Wakayama, Japan. *Tourism Management Perspectives*, *24*, 243–251. https://doi.org/10.1016/j.tmp.2017.07.017

Khoo-Lattimore, C. (2018). The ethics of excellence in tourism research: A reflexive analysis and implications for early career researchers. *Tourism Analysis*, *23*(2), 239–248. https://doi.org/10.3727/108354218X15210313504580

Kranzbühler, A. M., Kleijnen, M. H., Morgan, R. E., & Teerling, M. (2018). The multilevel nature of customer experience research: An integrative review and research agenda. *International Journal of Management Reviews*, *20*(2), 433–456. https://doi.org/10.1111/ijmr.12140

Lagos, E., Harris, A., & Sigala, M. (2015). Emotional language for image formation and market segmentation in dark tourism destinations: Findings from tour operators' websites promoting Gallipoli. *Tourismos*, *10*(2), 153–170.

Lennon & Foley. (2000). *Dark tourism: The attraction of death and disaster*. Thompson Learning.

Light, D. (2017). Progress in dark tourism and thanatourism research: An uneasy relationship with heritage tourism. *Tourism Management, 61*, 275–301. https://doi.org/10.1016/j.tourman.2017.01.011

MacCarthy, M. (2017). Consuming symbolism: Marketing D-Day and Normandy. *Journal of Heritage Tourism, 12* (2), 191–203. https://doi.org/10.1080/1743873X.2016.1174245

Magee, R., & Gilmore, A. (2015). Heritage site management: From dark tourism to transformative service experience. *The Service Industries Journal, 35*(15-16), 898–917. https://doi.org/10.1080/02642069.2015.1090980

Marcus, G. E. (1998). *Ethnography through thick and thin*. Princeton University Press.

Mkono, M. (2016). The reflexive tourist. *Annals of Tourism Research, 57*, 206–219. https://doi.org/10.1016/j.annals.2016.01.004

Mossberg, L. (2008). Extraordinary experiences through storytelling. *Scandinavian Journal of Hospitality and Tourism, 8*(3), 195–210. https://doi.org/10.1080/15022250802532443

Nawijn, J., Isaac, R. K., Liempt, A. V., & Gridnevskiy, K. (2016). Emotion clusters for concentration camp memorials. *Annals of Tourism Research, 61*, 244–247. https://doi.org/10.1016/j.annals.2016.09.005

Ngunjiri, F. W., Hernández, K. A. C., & Chang, H. (2010). Living autoethnography: Connecting life and research. *Journal of Research Practice, 6*(1), E1–E1. http://jrp.icaap.org/index.php/jrp/article/view/241/186

Norman, D. A. (2004). *Emotional design: Why we love (or hate) everyday things*. Basic Civitas Books.

Novelli, M. (2005). *Niche tourism: Contemporary issues, trends and cases*. Elsevier Butterworth-Heinemann.

Olmos-López, P., & Tusting, K. (2020). Autoethnography and the study of academic literacies: Exploring space, team research and mentoring. *Trabalhos em Linguística Aplicada, 59*(1), 264–295. https://doi.org/10.1590/010318136565715912020

Oren, G., Shani, A., & Poria, Y. (2021). Dialectical emotions in a dark heritage site: A study at the Auschwitz death camp. *Tourism Management, 82*. https://doi.org/10.1016/j.tourman.2020.104194

Packer, J., & Gill, C. (2017). Meaningful vacation experiences. In S. Filep, J. Laing, & M. Csikszentmihalyi (Eds.), *Positive tourism* (pp. 19–34). Oxon: Routledge.

Pine, II, B. J., & Gilmore, J. H. (1998). Welcome to the experience economy. *Harvard Business Review, 76*, 97–105.

Poria, Y., Reichel, A., & Biran, A. (2006). Heritage site management: Motivations and expectations. *Annals of Tourism Research, 33*(1), 162–178. https://doi.org/10.1016/j.annals.2005.08.001

Prayag, G., Hosany, S., Muskat, B., & Del Chiappa, G. (2017). Understanding the relationships between tourists' emotional experiences, perceived overall image, satisfaction, and intention to recommend. *Journal of Travel Research, 56*(1), 41–54. https://doi.org/10.1177/0047287515620567

Prayag, G., Hosany, S., & Odeh, K. (2013). The role of tourists' emotional experiences and satisfaction in understanding behavioral intentions. *Journal of Destination Marketing & Management, 2*(2), 118–127. https://doi.org/10.1016/j.jdmm.2013.05.001

Raine, R. (2013). A dark tourist spectrum. *International Journal of Culture, Tourism and Hospitality Research, 7*(3), 242–256. https://doi.org/10.1108/IJCTHR-05-2012-0037

Rose, G. (2016). *Visual methodologies: An introduction to researching with visual materials*. Sage Publications.

Sadala, M. L. A., & Adorno, R. D. C. F. (2002). Phenomenology as a method to investigate the experience lived: A perspective from Husserl and Merleau Ponty's thought. *Journal of Advanced Nursing, 37*(3), 282–293. https://doi.org/10.1046/j.1365-2648.2002.02071.x

Seaton, A. V. (1999). War and thanatourism: Waterloo 1815–1914. *Annals of Tourism Research, 26*(1), 130–158. https://doi.org/10.1016/S0160-7383(98)00057-7

Seaton, A. V. (2002). Thanatourism's final frontiers? Visits to cemeteries, churchyards and funerary sites as sacred and secular pilgrimage. *Tourism Recreation Research, 27*(2), 73–82. https://doi.org/10.1080/02508281.2002.11081223

Seligman, M. E., & Csikszentmihalyi, M. (2014). Positive psychology: An introduction. In M. Csikszentmihalyi (Ed.), *Flow and the foundations of positive psychology: The collected works of Mihaly Csikszentmihalyi* (pp. 279–298). Springer.

Servidio, R., & Ruffolo, I. (2016). Exploring the relationship between emotions and memorable tourism experiences through narratives. *Tourism Management Perspectives, 20*, 151–160. https://doi.org/10.1016/j.tmp.2016.07.010

Sharpley, R. (2005). Travels to the edge of darkness: Towards a typology of dark tourism. In C. Ryan, S. J. Page, & M. Aicken (Eds.), *Taking tourism to the limits: Issues, concepts and managerial perspectives* (pp. 217–228). Elsevier.

Sharpley, R. (2009). Shedding light on dark tourism: An introduction. In R. Sharpley & P. R. Stone (Eds.), *The darker side of travel: The theory and practice of dark tourism. Aspects of tourism* (pp. 3–22). Channel View Publications.

Shaver, P., Schwartz, J., Kirson, D., & O'Connor, C. (1987). Emotion knowledge: Further exploration of a prototype approach. *Journal of Personality and Social Psychology, 52*(6), 1061–1086. https://doi.org/10.1037/0022-3514.52.6.1061

Sidaway, J. D. (2002). Photography as geographical fieldwork. *Journal of Geography in Higher Education, 26*(1), 95–103. https://doi.org/10.1080/03098260120110395

Sigala, M. (2019). Scarecrows: An art exhibition at domaine Sigalas inspiring transformational wine tourism experiences. In M. Sigala & R. N. S. Robinson (Eds.), *Management and marketing of wine tourism business: Theory, practice and cases* (pp. 313–343). Palgrave Macmillan.

Slade, P. (2003). Gallipoli thanatourism: The meaning of Anzac. *Annals of Tourism Research, 30*(4), 779–794. https://doi.org/10.1016/S0160-7383(03)00025-2

Smit, B., & Melissen, F. (2018). *Sustainable customer experience design: Co-creating experiences in events, tourism and hospitality.* Routledge.

Smith, L. (2012). The cultural 'work' of tourism. In L. Smith, E. Waterton, & S. Watson (Eds.), *The cultural moment in tourism* (Vol. 26, p. 210). Routledge.

Spry, T. (2001). Performing autoethnography: An embodied methodological praxis. *Qualitative Inquiry, 7*(6), 706–732. https://doi.org/10.1177/107780040100700605

St-James, Y., Darveau, J., & Fortin, J. (2018). Immersion in film tourist experiences. *Journal of Travel & Tourism Marketing, 35*(3), 273–284. https://doi.org/10.1080/10548408.2017.1326362

The Statue of Liberty Ellis Island Foundation. (2020). *Ellis island history.* https://www.libertyellisfoundation.org/ellis-island-history

Stone, P. (2006). A dark tourism spectrum: Towards a typology of death and macabre related tourist sites, attractions and exhibitions. *Tourism: An Interdisciplinary International Journal, 54*(2), 145–160. https://hrcak.srce.hr/161464

Stone, P. (2012). Dark tourism and significant other death: Towards a model of mortality mediation. *Annals of Tourism Research, 39*(3), 1565–1587. https://doi.org/10.1016/j.annals.2012.04.007

Stone, P., Hartmann, R., Seaton, A., Sharpley, R., & White, L. (2018). *The Palgrave handbook of dark tourism studies.* Springer.

Straker, K., & Wrigley, C. (2015). The role of emotion in product, service and business model design. *Journal of Entrepreneurship, Management and Innovation, 11*(1), 11–28. https://doi.org/10.7341/20151112

Straker, K., & Wrigley, C. (2016). Translating emotional insights into digital channel designs. *Journal of Hospitality and Tourism Technology, 7*(2), 135–157. https://doi.org/10.1108/JHTT-11-2015-0041

Straker, K., & Wrigley, C. (2018). Engaging passengers across digital channels: An international study of 100 airports. *Journal of Hospitality and Tourism Management, 34*, 82–92. https://doi.org/10.1016/j.jhtm.2018.01.001

Timoney, S. (2019). We should all know where we came from. Identity and personal experiences at heritage sites. *Journal of Heritage Tourism,* 1–14. https://doi.org/10.1080/1743873X.2019.1667999

Vagle, M. D. (2018). *Crafting phenomenological research* (2nd ed.). Routledge.

Van Manen, M. (2016). *Phenomenology of practice: Meaning-giving methods in phenomenological research and writing.* Routledge.

White, L., & Frew, E. (2013). *Dark tourism and place identity: Managing and interpreting dark places.* Routledge.

Winter, C. (2011). Battlefield visitor motivations: Explorations in the great war town of Ieper, Belgium. *International Journal of Tourism Research, 13*(2), 164–176. https://doi.org/10.1002/jtr.806https://doi.org/10.1002/jtr.806

Wrigley, C., & Straker, K. (2019). *Affected: Emotionally engaging customers in the digital age.* John Wiley & Sons.

Zheng, C., Zhang, J., Qiu, M., Guo, Y., & Zhang, H. (2020). From mixed emotional experience to spiritual meaning: Learning in dark tourism places. *Tourism Geographies, 22*(1), 105–126. https://doi.org/10.1080/14616688.2019.1618903

Designing dark tourism experiences: an exploration of edutainment interpretation at lighter dark visitor attractions

Brianna Wyatt ⓘ, Anna Leask ⓘ and Paul Barron ⓘ

ABSTRACT

Existing dark tourism literature has explored various aspects of interpretation, including challenges in balancing interpretation efforts with concerns for historical accuracy, and managing ethical issues with interpreting past tragedies for packaged tourism purposes. However, research appears under-developed concerning the influences on the design of interpretation at dark visitor attractions, particularly those considered *lighter* due to their edutainment agenda. This paper thus critically explores the influences on the design of edutainment interpretation at three lighter dark visitor attractions, which are introduced as new attractions for study within dark tourism research. It also discusses the findings achieved that not only contribute to the study's conclusions and recommendations for future research in the realms of dark tourism and interpretation, but also contribute to enhancing interpretation design understanding for both dark tourism research and practice.

Introduction

Dark tourism is generally referred to as an act of travel to sites associated with death, suffering, and the seemingly macabre (Stone, 2006, p. 146). Now a widespread and diverse area within the tourism industry (Hooper, 2017), dark tourism has become used as an analytical lens to promote academic discussion relating to interpretation and issues of mixing leisure and entertainment with commemoration and tragedy (Dunkley, 2017). Yet, research remains challenged by the range of interdisciplinary studies that offer divergent perspectives on the management and operations of dark tourism activities, specifically interpretation (Jamal & Lelo, 2011; Stone, 2013).

Supporting the topic of this special issue and the broader exploration of how dark tourism experiences are created, this paper explores the influences on the design of edutainment interpretation within dark tourism. Although past publications have addressed interpretation within dark tourism and the issues and challenges underpinning it, much of these publications have largely focused on dark visitor attractions (DVAs) of the darkest nature– those that represent modern tragedies through commemorative and educational agendas (Wyatt, 2019). The preference given to darker DVAs has consequently developed a lack of attention given to DVAs situated at the lighter end of Stone's (2006) Darkness Spectrum. These lighter DVAs, or rather LDVAs, are recognised for their higher tourism infrastructure and commercially driven, edutainment agenda– an interpretation approach that uses innovative and engaging methods to create experiences that are both educational and entertaining (Wyatt, 2019). Although edutainment has been criticised for its use within dark tourism, some studies have demonstrated an overwhelming visitor preference for mixing education with fun in dark tourism experiences (Ivanova & Light, 2018). Thus, by exploring the influences on the design of edutainment interpretation within dark tourism, this paper enhances dark tourism research and understanding relating to the diversity of LDVAs, their interpretation, and the use of edutainment agendas within dark tourism.

Dark tourism, interpretation and edutainment

Dark tourism is a highly complex and multidimensional phenomenon involving visits to real and recreated places associated with death, suffering, misfortune, and the seemingly macabre (Fonseca et al., 2016; Stone, 2006). Growing exponentially over the last twenty-five years, dark tourism has become an increasingly significant component of the wider heritage tourism industry. However, noting its perceived exploitation and trivialisation of historic tragedies, scholars have suggested dark tourism is, in practice, beset with ethical and management challenges (Dalton, 2015; Foley & Lennon, 1996). Still, as a catalyst for emotional values and knowledge enrichment (Kim & Butler, 2015), dark tourism can offer audiences opportunities to connect with difficult pasts through interpretation.

As an informational and inspirational process designed to enhance understanding, appreciation and conservation of heritage assets (Beck & Cable, 2002), interpretation is an essential component for all visitor experiences. Grounded in the art of storytelling, Smith (2016) argues interpretation is charged with the task of enhancing visitor understanding through thought-provoking displays that encourage visitors to be less passive in their visits. Key to interpretation, and subsequently the overall visitor experience, is the interpretation design (Roberts, 2014), which communicates interpretive plans into a tangible form (Woodward, 2009). Within dark tourism, interpretation offers visitors emotional, educational and/or entertaining experiences, where they can connect their memories, knowledge, and interests with the history and heritage on display (Kavanagh, 1996). Although some dark tourism experiences are morally contested, the ultimate goal of their interpretation design is to communicate the significance and meaning of heritage to visitors (Grimwade & Carter, 2000), and allow them to utilise their understandings of the past in order to make sense of their visitor experience (Kidd, 2011).

The discourse concerning dark tourism experiences has led to an acknowledgement of the wide range of DVAs. As the physical manifestations of historic death and tragedy, whether in-situ or purposefully constructed, DVAs often occur as a result of the intentional exploitation of dark heritage through tourism activities underpinned by a strategically designed interpretation (Tarlow, 2005). In highlighting the fact that DVAs often render ideological agendas that are intertwined with interpretation and meaning, Stone (2018) suggests they help to reveal the idiosyncrasies of social histories that can provoke feelings of anxiety, remorse, empathy, or fear. Yet, as large memory vessels, DVAs can provide cathartic, commemorative, or educational experiences that help society to cope with past tragedies, as well as endorse feelings of shock, thrill, and even enjoyment through edutainment agendas. Although the use of edutainment within dark tourism is a contentious topic, Santonen and Faber (2015) suggest it can provide benefits of increased visitor motivation, retention, and active learning. Because of this, edutainment has become a preferred technique within the wider heritage tourism industry.

As the amalgamation of educational and cultural activities with the commercialisation and technology of the entertainment world (Hannigan, 1998), edutainment is fundamentally grounded in the notion that learning can be fun. It has been further described as both 'entertainment that educates' and 'education that informs and entertains' (Ron & Timothy, 2013). This concept of linking education with entertainment is traditionally associated with the work of Walt Disney. In observing existing amusement parks as meaningless, outmoded, and lacking any form of educational contribution, Disney developed and popularised the practice of theming (Oren & Shani, 2012), which Åstrøm (2020) suggests is an extension to set design and may be viewed as a staging process that unifies education, entertainment and technology through strategic organisation and structure. Thus, if edutainment is the agenda, then theming is the method.

Scholars have suggested the commercial success of Disney's themed environments and the need to overcome increasing challenges of art shock and visitor fatigue led museums to become the first heritage tourism sector to adopt theming and assimilate education with entertainment in practice (Hannigan, 1998; Hertzman, 2006; King, 1991). As a less didactic and informal method, theming, coupled with storytelling and technology, has been observed as an effective way to focus visitor attention, while maintaining their interest, in order to foster a deeper learning experience (Oren & Shani, 2012; Ron & Timothy, 2013). Since then, theming has been shown to provide added value to an array of visitor experiences (Hannigan, 1998), further contributing to the transformation of many heritage attractions into ideal edutainment tourism products (Hertzman et al., 2008). Thus, Disney's legacy in relation to other sectors, including dark tourism, is his incorporation of storytelling, technological advancements, education and entertainment, customer service, and a recognition of consumers' changing consumption habits (Shani & Logan, 2010).

Despite the success of edutainment within the wider heritage tourism industry, much of dark tourism research criticises it and the 'Disney Effect' that has infiltrated dark tourism practice (Dalton, 2015; Isaac & Çakmak, 2014; Stone, 2009a). Although some scholars have questioned whether edutainment is a sufficient and/or appropriate form of interpretation (Dunkley, 2017; Hooper, 2017), others have argued it provides a framework that, through effective thematic storytelling, helps to create meaningful experiences that resonate with visitors even after they leave a location (Heidelberg, 2015; Oren & Shani, 2012). Regardless of research opinions, dark tourism edutainment experiences have risen in popularity in recent years (Heidelberg, 2015) with an increasing number of attractions and tours, such as the Dungeons experiences and commercial ghost tours. The popularity of these attractions is arguably connected to the commercialisation of dark tourism themes, despite the macabre undertones (Bristow & Jenkins, 2020). Still, the growing interest in seeking out scary experiences for pleasure has developed an expanding market of LDVA attractions, and thus becoming an international phenomenon (Bristow & Jenkins, 2020; Holloway, 2010).

Generally associated with LDVAs, dark tourism edutainment experiences are delivered through either a heritage-centric approach that seeks to educate and create appreciation of dark heritage (e.g. Eden Camp, Tallinn Legends, Gettysburg, Colonial Williamsburg), or a fun-centric approach that seeks to shock and thrill audiences (e.g. Edinburgh Dungeons, London Ghost Bus Tour, Jack the Ripper Tour). Both approaches rely on theming and storytelling in order to create an understanding of the history among visitors (Ron & Timothy, 2013). However, research has demonstrated LDVAs are largely criticised as insignificant amusements that sanitise historical truths through narrative softening and omission and whitewashing the environment (Dwyer & Alderman, 2008; Silverman, 2011; Stone, 2006). In response, others have argued that through edutainment, LDVAs actually educate visitors and fulfil curiosities about darker histories of the more distant past through raw and realistic representations (Magee & Gilmore, 2015; Rodriguez-Garcia, 2012; Welch, 2016).

Despite the academic interest, both research and understanding are under-developed in relation to the range of edutainment experiences within dark tourism, or rather the diversity of LDVAs, and how these experiences are designed and managed over time. This is predominantly due to the issue that few publications have explored LDVAs (e.g. Gentry, 2007; Holloway, 2010; Ivanova & Light, 2018; McEvoy, 2016; Powell & Iankova, 2016; Rodriguez-Garcia, 2012), which is largely a consequence of the continued research attention directed towards the darkest forms of DVAs. The reasoning for this oversight is unclear. However, Ivanova and Light (2018) suggest it may be that scholars perceive the darkest DVAs as more deserving of academic scrutiny, as they raise broader questions relating to commodification and authenticity when compared to LDVAs. Yet, considering the critical views and discourse that surrounds LDVAs, it would be reasonable for research to give them greater attention in order to answer questions relating to their commodification and authenticity.

Further complicating the under-development of LDVA research, past publications have largely relied on ghost tour experiences, the London Dungeons, and Jack the Ripper tours, consequently discouraging the recognition of LDVA diversity. What is more, research on LDVAs and their edutainment interpretation has become fragmented by studies that have explored individual underpinning nuances of interpretation, such as selectivity and narrative development (Spaul & Wilbert, 2017; Watson, 2018); exhibition presentation (Rátz, 2006; Wight & Lennon, 2007); issues with authenticity (Heuermann & Chhabra, 2014); and the role of tour guides in interpreting sensitive histories (Potter, 2016; Quinn & Ryan, 2016). Although these publications have contributed to an awareness of interpretation within dark tourism, they largely report on specific methods and ethical concerns. As a result, the influences and practical processes of interpretation or designing an edutainment experience for LDVAs remains under-analysed. Given the fact that interpretation design blurs the boundaries of exhibition, object display, and the visitor environment, resulting in an immersive and multisensory experience (Roberts, 2015; Woodward, 2009), it would seem important for dark tourism research to create a better understanding for how and why an experience is designed in the way that it is. In doing so, new light may be shed on the LDVA experience, thereby demonstrating how these products support visitor understanding and meaningful experiences.

Methodology and methods

Underpinned by an interpretative perspective, this qualitative, exploratory research was grounded in the subjects of heritage tourism, dark tourism, and interpretation in order to expand knowledge on the design of edutainment interpretation within dark tourism. In particular, it used three LDVAs that were associated with pre-nineteenth century history, specifically the plague– a biological disaster, dubbed 'the great mortality', that swept across Europe from the mid-fourteenth to the late-seventeenth century (Platt, 2014), claiming to have killed more people than any other single known historical event (Beaumont, 2014) and is exceeded only by WWII in terms of devastation, human suffering and loss of life (Kelly, 2006). These criteria were selected because when compared to other historical tragedies, research remains underdeveloped in relation to the interpretation of pre-nineteenth century tragedies (e.g. plague, Medieval torture, crime and punishment, persecution, witch burning, etc.). Although the London Dungeon and ghost tours have been the focus of previous publications (see e.g. Hovi, 2008; Ivanova & Light, 2018; Stone, 2009), these studies are predominantly visitor-focused, or are set within the wider realm of heritage tourism studies.

Data collection was completed at three different LDVAs, identified through purposive sampling– The Real Mary King's Close (RMKC) in Edinburgh, the Sick to Death museum (S2D) in Chester, and the Gravedigger Ghost Tour (GGT) in Dublin. As an in-situ attraction (RMKC), a static museum (S2D), and a bus tour (GGT), these LDVAs were used as example representations of the wider range of LDVAs not yet explored in dark tourism research. They each presented an interpretation design using edutainment, which aimed to educate, provoke and engage audiences through a variety of methods (e.g. self-guided tours, character re-enactment, exhibitions, set dressings, innovative technologies) about the harsh realities of sixteenth – eighteenth century life, and, in particular, the plague. These three LDVAs were used since, to date, they had not previously been used in research. In addition, three LDVAs were used as the literature has been widely conducted through descriptive, and often single, case study approaches (Ashworth & Page, 2011; Ioannides et al., 2014; Leask, 2016).

Data were collected using semi-structured interviews and focus groups, of which informed consent for participation and use of job titles was confirmed and obtained. These methods were selected given their preferred use in dark tourism research (Light, 2017; Wight, 2006) and also, that they often lead to a breadth of commentary and analysis, as well as opportunities for transferability (Goulding & Domic, 2009; Korstanje, 2018). Yet, in order to shed new light on current understanding and contribute to existing discourse, rich picture building (RPB) was also used during the focus group sessions. RPB, illustrated in Figure 1, is a data collecting tool used during focus group sessions to help develop discussion and aid participants in expressing, through pictorial representation, their emotions, perceptions, and conflicted understandings about a topic (Ho, 2015). Thus, RPB is seen as a beneficial tool for evoking and recording insight into social situations (Bell & Morse, 2013b). Although RPB can be weakened by a lacking central theme, an overabundance of written words, and inadequate use of colour, its benefit of aiding problem solving and creative thinking has led it to become a useful technique, particularly in the social and behavioural sciences, because humans are thought to communicate more easily through impressions and symbols than words (Bell & Morse, 2013b). Although a tool for exploring social issues and situations, RPB has not previously been used in dark tourism research. Thus, its introduction and use in this study is a key contribution to both dark tourism research and practice.

Data collection was carried out through twelve semi-structured interviews across the three LDVAs and included managers and designers that were involved with the design of the interpretation. The interview questions were drawn from the literature that discussed interpretation and design processes. According to the literature, influences on interpretation can include stakeholder roles; experience with designing interpretation; personal preferences; space; access limitations; authenticity concerns; conservation; the budget; and timeframe (Brochu, 2003; Jones, 2007; Knudsen et al., 1995; Roberts, 2015).

Focus group sessions using RPB were also conducted at each LDVA with staff/guides. The decision to include the staff/guides in the data collection was based on the literature, which argued they are the mediators of meaning and the interface between an attraction and its visitors, charged

Figure 1. Rich Picture.

with the responsibility of promoting the interpretation (Bryon, 2012). Due to the LDVAs' size and time in operation, two focus groups, each of five guides, were conducted at RMKC, while one focus group of three guides was conducted at GGT and one focus group of two staff members was conducted at S2D. The focus groups were asked to consider the prompt: *What is your perception of the design and management of interpretation at your attraction?* The prompt was derived from the literature, which suggested guides are constantly making judgments about how an interpretation design is working for audiences (Potter, 2016). The staff/guides were therefore able to provide first-hand perspectives on how the interpretation design was working for the visitor experience. Using coloured markers and poster paper, the participants were asked to consider the prompt and then collectively draw, in pictorial form, their personal perspectives and opinions about their attraction's interpretation, how it was designed, how it is currently managed, challenges because of the design, and any solutions for overcoming those challenges. These sessions also included group discussions that further contributed to the pictorial representations.

The findings of this research were analysed through thematic analysis, which helps to identify, analyse and interpret themes within collected qualitative data, and capture relevant and significant meaning in the data (Clarke & Braun, 2017). The interview and focus group recordings were transcribed by the author and manually analysed using line-by-line coding in relation to their corresponding LDVA, considering each maintains a distinct purpose, thereby employing different methods for interpretation. The analysis of the rich pictures followed Bell and Morse's (2013a) suggestion of applying Carney's (1994) seven-step process for critiquing art. The decision to use this analysis method is grounded in the belief that it merges formal analysis with interpretation (Bell & Morse, 2013a). By looking at the stylistic features of the pictures, one can better understand the participants' aims and goals, what the participants put into the overall picture and therefore deem as important, and the relationship of the drawings to past and future issues (Carney, 1994). Thus, the pictures were analysed for both their content and context, which allowed for an interpretation of meaning, revealing that the pictures were not narrowly focused, but rather encompassed numerous topics underpinning the LDVAs' businesses, which impact or are impacted by the design and management of the interpretation.

Findings and discussion: influences on edutainment interpretation

To add to the influences earlier mentioned and as discussed in the following, three additional factors were identified as being influential on the LDVAs' edutainment interpretation: pop-culture references, the nature of the content, and other attractions and competition.

Pop-culture references

The findings for this study revealed several pop-culture references were highly influential on the LDVAs' edutainment designs. These include horror movies of the 1980s and the book series and TV show, *Horrible Histories*, which, aimed at 8+ year olds (Berenbroek, 2013), pushes moral boundaries in its blending of satirical imagery and performances with horrible and unfortunate histories of various historical periods from the Egyptians to the Victorians (Scanlon, 2011). Challenging the elitist nature of school history programmes that tend to overlook the lives of ordinary people, *Horrible Histories* is a commentary on the way in which history has been remembered, further arguing that history is merely a version of someone's perspective (Scanlon, 2011). This attitude is shared by S2D's Director, who explained that when designing S2D's interpretation, he wanted to break from the conformities of high-brow institutional thinking and create an interpretation that was unapologetic in its delivery of raw and provoking displays. Inspired by the *Horrible Histories* use of irreverent humour and preoccupation with unpleasant and gory historical accounts (Scanlon, 2011), S2D's Director defined the experience as a union of science, history, and the fun style of *Horrible Histories*. S2D's interpretation was thus designed to encompass gross-out factors, sensory stimulation and hands-on activities that promoted informal learning experiences. Some visitors may be shocked by the image of a disembowelled hanging man used to reflect the internal effects of the plague, or the smells and sounds of a man suffering from either dysentery or intestinal worms. However, the Director explained,

> When people are shocked by the displays, they're usually like 'what is this about?' Here, we provoke learning. We're not going to *not* do something just because people might be offended.

(S2D, Director)

This explanation of provoking visitors is underpinned with an intention to provoke curiosity, questions, and subsequently learning. The Director explained that some of the displays are shocking, but in a fun-science way. For example, he described an initial exhibit of a giant anus that visitors could remove fistulas from while learning about diseases and complications of the bowel. He also had plans to create a green screen exhibit where visitors could place themselves into a burial procession while learning about changes in burial processes throughout history.

In their focus group, the S2D staff suggested the museum was saturated with learning experiences and could actually do more to enhance the entertainment side of the design. Drawing a plague doctor, as illustrated in Figure 2, one staff member argued that most visitors ask why there are no character actors. In fact, it was agreed that most visitors seem disappointed with the experience because of this. When asked why S2D does not use character actors, the Head of Operations explained it was due to a lack of space, staff and funding. However, they do use a plague doctor and Medieval surgeon actors for school groups or during special events. In place of actors, another staff member suggested the museum could have more hands-on activities and sensory stimulation to enhance the experience.

Figure 2. S2D Rich Picture.

Comparably, GGT's tour was also influenced by the *Horrible Histories* series. According to the Manager, the delivery of the tour's narrative was designed to be very much tongue-in-cheek, similar to the *Horrible Histories* satirical style. The Manager commented that they always get people on the tour expecting it to be scary because of how the bus looks and instead they find themselves laughing at a foul-mouthed plague victim spitting in their hair and making boorish jokes. The staff echoed this, stating that they often get people leaving saying that they didn't expect they were going to laugh so much. Thus, the staff described the tour as a theatre on wheels, poking fun at Dublin's suffering from the plague's devastation, along with other macabre histories. Commenting on the tour's treatment of history and use of terrible toilet humour, the Manager explained that the narrative is based on facts drawn from thorough research into Dublin's macabre history. The Manager further explained that while the tour's interpretation is heavily driven by entertainment values, it does seek to educate visitors about the macabre history of Dublin. This echoes the informal learning objectives of the *Horrible Histories* series, which further demonstrates the close associations between it and GGT.

The majority of the GGT tour relies on the performance abilities of the actors, while the bus and its aesthetic design are themed under the inspiration of schlocky horror movies from the 1980s. Equipped with fake skeletons, ambient lighting and stylised hard foam, the lower deck of GGT's bus was designed to resemble a crypt, which visitors must pass through to reach the top deck that is fashioned with blackout curtains, coloured lights, strobe effects, and fake bones for handrails. Both the Manager and Designer emphasised the significance of the film *Evil Dead* as having been influential on the entire tour, including the style of acting and makeup. Speaking of the overall feeling and atmosphere for the tour, they explained that the film and score of *Jaws* was highly influential. Explaining the significance of sound design, the Designer stated,

> Sound design is a huge thing. Look at something like *Jaws*– it is cool, certainly not as scary. But that iconic sound builds the tension.

(GGT, Designer)

Creating the right atmosphere and building the tension was incredibly important for GGT. One staff member explained many people on their tour do not speak English fluently and can have a hard time understanding their accent. Therefore, the actors often have to rely on their own animation and the special effects to help build tension and create a memorable experience.

On the topic of building tension, the Manager explained that in his experience he had found no one wants to be scared for too long. Referring again to horror movies, he explained there are points in movies where the audience laughs and points where they are scared, but the aim is to find the right balance. This was echoed in the focus group, where the actors suggested it is not about doing too much of either– comedy or scare. Rather, it is about finding the right moment to scare. They explained,

> The whole thing is very basic when you think about in terms of scare attractions. Everyone uses the jump scare, and they use it because it works, even in cinema. But this sort of 'easy scare' isn't really all that easy because it is all about timing and putting people in the right atmosphere, and then it is funny.

(GGT, Actor 1)

> It is all about making people feel at ease, making it funny, and then scaring them. Or making them feel uneasy and not scaring them, and then it is just the effect. I use the example of creating the atmosphere like in a movie where it's really intense, but then the scare is just a cat.

(GGT, Actor 2)

The actors' discussions about the effect of jump scares corresponds with the thinking of early horror films, such as *Psycho*. As Alfred Hitchcock once said, there is no terror in the bang, only the anticipation of it (Skov & Andersen, 2001). Yet, as the tour was not intended to be overly scary, there was an importance in finding the right balance between scare and comedy. The actors explained, in order to make it fun and entertaining, and to prevent mass hysteria, there needs to be an element of relief through comedy and satire.

The nature of the content

Understandably, the findings also revealed that the nature of the content was an influence on the LDVAs edutainment interpretation. Unlike darker DVAs that require a sense of gravity in their retelling of more modern tragedies, the temporal distance between the present day and the historical events of the plague have allowed for the LDVAs to take on a more light-hearted approach. On this topic, GGT's Manager explained their use of comedy for this history is justified because,

> It was a shared tragedy and it happened absolutely everywhere. To be honest, most people's ways of dealing with something horrific is to joke about it. Sometimes the only way you can engage people is to disguise it as entertainment, so they learn something without realising they've actually learned.

(GGT, Manager)

It is apparent that there were no concerns relating to the nature of the content and the need to be sensitive in its delivery. GGT's Manager supported this by explaining they were not creating a museum piece. Therefore, there was no obligation to be sensitive towards the history. The edutainment purpose of the tour is the reason why it was designed like a set for a horror movie. To further elaborate, one actor explained that the main character is a nameless plague victim, removing any onus to someone's memory. On developing the main character of the tour, he commented,

> The plague was spread over hundreds of years. We did research, but because it wasn't specific to a period, we weren't dealing with a specific character or a specific time, it opened the character up to be whatever we wanted.

(GGT, Actor)

Yet, the Manager stated they were careful not to push certain boundaries too far. Acknowledging that they talk about horrific events, the Manager explained that they do it in a way that is not necessarily horrific. He commented,

> We are not gory for the sense of gore, or for shock tactics. People are perfectly capable of going to that place themselves. It's not necessary. Some of the stories we tell, like of the Dolocher, are horrific, but we tell them in such a way that you get the story. It would be so easy to be gory with things that are horrendous, but we don't. We let it sit and you can go as far into it as you want.

(GGT, Manager)

Comparably, the nature of the content was also an influence on RMKC's interpretation, and in similar fashion, the design was not created for shock tactics. Rather, the CEO explained that prior to coming under the ownership of Continuum Attractions, RMKC had been operating as a ghost tour that claimed people with the plague had been bricked up and left to die. Research proved this to be untrue; and wanting to correct the mangled history in order to create a fact-based experience that would educate their guests about the real history of life on the Closes, their focus became presenting a story that would deliver accurate information while debunking the myths. Still, recognising the commercial viability of popular public assumptions relating to the ghostly atmosphere of RMKC, the CEO stated,

> We are not supported by external funding. Therefore, we do use terms like 'deep beneath the streets' or 'deep beneath your feet.' The script is factually correct, but we can't get away from people leaving thinking it was spooky. So, are we playing to it? We are using it to get people in, but we are certainly not deceiving them when they come through.

(RMKC, CEO)

Contributing to this discussion, the Head of Development explained that when taking on RMKC, they knew they had a set of stories that could be difficult to handle due to the nature of the history. However, they maintained the perspective that it was a just part of history. Therefore, certain elements that they added to the physical design, like the vomit bucket in the plague room, which was fitted with a smell pod that emits a vomit scent, was not meant to create a darker or more horrific experience. Rather, it was to create a real and fact-based experience that was true to the history. In agreement, the General Manager explained,

> The facts about Edinburgh are scarier than the fiction that people make up. We are telling Edinburgh's story, and Edinburgh's story is a lot more intriguing, dark and uncomfortable than any of the best Stephen King or Clive Barker stories. We bring that to life.

(RMKC, General Manager)

It is clear that the narrative was written to deliver fact-based information, and according to the Designer and General Manager, less was more in terms of the set dressing. The furniture and props used were researched for likeness to what would have been used in the sixteenth and seventeenth centuries. The Designer explained they relied on the ambient lighting and sensory effects to help bring the story to life.

However, in discussing the added elements of lights, the staff voiced concerns over the placement of some lights and the use of coloured lights. For example, one staff member stated,

> I understand the intent of the coloured lights. We have a green light on in the plague room because it is not a very pleasant colour and it was this horrible disease. But at the same time, it is a historical tour and they wouldn't have had this bright, blinding green light in the 1600s. It's distracting. People have to find the right position to block their eyes from it. It is taking away from the experience.

(RMKC, Guide)

The staff continued to explain that on occasion the lights will malfunction and circulate through all the colour options, creating a disco-like light show for the visitors. This is not appropriate considering the nature of the content. Further commenting on the satirical elements of the script, one guide stated,

> The plague killed half the people on the planet at one point. But because cognitive distance is created, because of the historical barriers and media, people have become desensitised. So, we have mock-up models of people dying, people covered in boils, one is a child, and one is a baby. But it's people dying.

(RMKC, Guide)

In discussing the nature of the content, one guide stated that they are telling guests how people used to live and genuine things that happened to them, but because of the desensitised perspective that visitors bring with them, they are often left underwhelmed by the lack of scare and shock tactics. Yet, in discussing their rich picture, another guide commented that they think the interpretation is actually airbrushed, or softened, and that parts of the script could be more descriptive to really show how horrible this time in history really was.

Equally, S2D's interpretation was also influenced by the nature of the content and further managed by finding a balance between it and its delivery. Although the Director talked about adding more gore and blood in future, he explained that it is always grounded in fact-based science in order to educate their visitors. The Director commented that during the planning stages there was apprehension among the other managers about some of the graphic details and how much blood and guts they were using. However, he argued that he was not bothered with pushing the envelope.

This desensitised perspective of the nature of the content and how it is delivered through the interpretation's design does not only appear shared across all three LDVAs, but S2D's Director explained he had recognised a morbid fascination among the public that seems associated with the concept of mortality. When asked if he was concerned about the nature of the content and motivations for why visitors wanted to see blood and gore, he stated,

> I have no issues at all with people having a morbid fascination with the body, death, and dying. Morbid curiosity is just as good as any curiosity. Our objective is to get people to learn or become interested in the history. We certainly want learning, having fun, and enjoyment, and if morbid curiosity is something people are interested in, then that's fine as well.

(S2D, Director)

The emphasis on science, medicine and education was reiterated by S2D's Head of Operations, who, like the Director, dismissed any concerns for needing to be sensitive in relation to the delivery of the content. In discussing some of the physical features of the design, she explained that they wanted to have a lot of hands on activities. Visitors are able to touch replica body parts afflicted by various diseases, take pictures with leprosy victims, observe the plague under a microscope in a replica plague house, and touch human skulls impacted by disease and trauma. In creating these activities, the Head of Operations stated that their attitude towards the nature of the content was that it was just history and science. Some people might think it is gory, but in reality, it is what happened. She explained,

> There is nothing wrong with discussing the realities of Medieval life. We wanted to say, 'this is what it would have looked like.' There are certainly the gory elements of it, for example the hanging man. You could maybe say we took a strong approach to that, but we do make people aware of it before they turn up so they can make their choice to come in or not come in.

(S2D, Head of Operations)

For this reason, the Content Expert stated that in comparison to other museums that recreate this time period, S2D is certainly more upfront about the more unpleasant aspects. She further revealed that the light-hearted *Horrible Histories* approach seems to help with delivering the heavier content that can sometimes be difficult to cope with. As her role was to ensure S2D was not playing into the myths of the plague that had been created in film and mass media, the Content Expert explained that the overall message being conveyed was that Medieval life was hard and people of that time were far less fortunate than we are today, particularly in terms of treating diseases, such as the plague.

Other attractions and competition

Given the fact that LDVAs maintain a higher tourism infrastructure and are commercially driven, it is logical that the interpretation of LDVAs for this study were influenced by other attractions in the market. On this topic, the Head of Development for RMKC explained how their interpretation was influenced by Jorvik Viking Village, Canterbury Tales, and the York Chocolate Story. Discussing the nature of the content and how they considered its delivery, he explained from their experience in creating the 'Eric the Bloodaxe' display for Jorvik Viking Village, they knew that the darker stories and slightly more graphic designs appealed to people. It was for this reason that the plague became a focal point for RMKC. However, because RMKC is located beneath the streets of Edinburgh, and as the plague is a primary focus, there were concerns that RMKC would be viewed as a dungeon experience. To avoid this, RMKC was contractually obligated by the City Council not tell ghost stories or the same stories that were being delivered in the Edinburgh Dungeons. In addition, the script was written to include a history of why and how RMKC came to be underground.

It is important to note the difference between RMKC and the Edinburgh Dungeons, because it was revealed to have been an influence of what RMKC did not want to do with their interpretation. According to the General Manager, unlike other attractions in the city, RMKC tells the real story of Edinburgh and life on the Closes in the sixteenth and seventeenth centuries without the support of ghost stories and scare factors. Describing the Dungeons as an adrenaline experience, he explained

RMKC is different because they rely on the nature of the story to scare people. He further commented,

> The adrenaline that we create is naturally created from peoples' apprehensions, perceptions, and expectations. Going back to plague room, visitors will see the silhouette of the plague doctor and make an assumption that someone is going to jump out and scare them. There is a heightened adrenaline, and then that doesn't happen. The adrenaline drops and then the chemical imbalance in their body happens and they feel a bit queasy or faint, why? Because they got themselves worked up. We don't deliberately do that.

(RMKC, General Manager)

Supporting the General Manager's statements, the Guiding Manager explained that RMKC tries to keep the interpretation on the lighter side, like the Dungeons. However, he argues the Dungeons is more about historical passion with entertainment and lacks the same heritage element that RMKC holds, which sets them apart. He explained RMKC does have an entertainment value, but it is more educational than entertaining.

In discussing RMKC's entertainment features, the use of smell pods to create a sensory experience was also inspired from their work with Jorvik, where this technique had been successful. Moreover, the IT Manager explained that the decision to use character actors to deliver the tour was an idea that had developed from the Canterbury Tales attraction– an interactive tour of Geoffrey Chaucer's stories, which has now permanently closed. According to the Head of Development, the idea was,

> When you have something that has been well-received, don't re-invent the wheel, just replicate it and work it around the different storyline.

(RMKC, Head of Development)

However, referring to their rich picture, as illustrated in Figure 3, and in discussing the use of character actors, the guides expressed concerns for the sense of authenticity of both RMKC's history and the site. One guide commented that the costumes do have an element of Disney and that the tour would feel more like a heritage tour if the guides were in a standard uniform. Discussing how the guides are meant to act like characters from the 1600s, one guide commented,

> There is a strong focus on 'don't make modern day references', 'when you're outside, you are in character', 'don't forget when you're in costume people think…' No, they don't. People do not think that! They

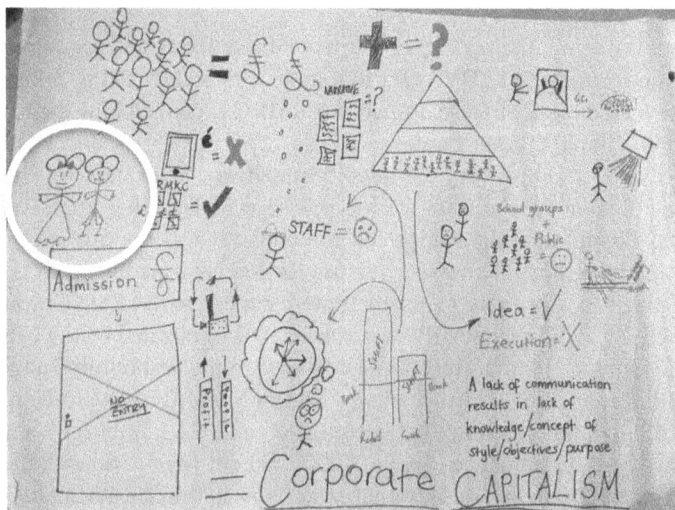

Figure 3. RMKC Group 1 Rich Picture.

don't see a regular person dressed in slightly odd clothing and think 'oh, they are from the past.' We are not Disneyland.

(RMKC, Guide)

Despite these perspectives, there seemed to be an agreement among the guides that if they were to continue with the character actors, there was a need for more engagement between them and the visitors, as well as more actors on site to recreate different stories and enhance the overall experience.

It was revealed that RMKC's interpretation was further influenced by the York Chocolate Story, another attraction owned by Continuum Attractions, which uses the technology of talking portraits seen in the *Harry Potter* movies. Considering the success and positive visitor feedback for the portraits, talking portraits of Mary King, Dr. Arnett and Sir Archibald Johnston of Warriston were used to replace the original gobo lights in RMKC's Gallery Room for RMKC's 10th anniversary. Subsequently, with regards to this particular feature, RMKC's interpretation was influenced by the York Chocolate Story and, indirectly, by the *Harry Potter* movies.

It was also revealed that S2D and GGT were influenced by other attractions. While the concept of GGT's tour came from the Manager's experience working with an on-site theatre tour that took place on location of the stories told, S2D's Head of Operations revealed most of their displays and exhibits were copied from various attractions in York. Specifically, the exhibits on blood-letting, urine analysis, herbal remedies, and some of the hands-on activities were copied from the Richard III museum and amalgamated into the S2D experience.

What is more, discussing conceptual developments of S2D's interpretation, both the Director and Head of Operations referred to the London Dungeons as a form of inspiration in the sense of shock factors. The Head of Operations explained that some of the gory interactive features and entertaining narratives are very much like what is being done at the Dungeons. However, the Director was explicit in stating that S2D is much more academically grounded, as it is focused on educating with elements of entertainment.

Conclusions

The findings present a clear understanding that although influenced by pop-culture references and promoting edutainment agendas, the LDVAs each sought to design their interpretation in a manner that would educate their audiences and provide historically accurate and academically grounded information. Despite the temporal distance of the history, higher tourism infrastructure, and commercial elements, there was a recognition among the LDVAs of the importance in providing a fact-based experience. Through raw and real recreations of the history, the LDVAs sought to provoke visitor learning and engagement. It was further demonstrated that none of the LDVAs sought to embellish the history or create gore for the sake of gore. These findings are suggested to be the result of a desensitised perspective among the LDVAs' management teams for the nature of the history. However, as demonstrated in the focus groups, there was a common understanding that most visitors share this desensitised perspective as a result of film and mass media, which has consequently led to their expectations of greater scare and shock factors. Although the goal to deliver fact-based experiences was challenged by commercial concerns to maintain business, there was a shared understanding among the LDVAs' management and staff that the history was priority and finding the right balance between the education and entertainment was essential.

To conclude, in contrast to current understanding of LDVAs as being unconcerned with matters of facts and historical accuracy, this study has demonstrated they are in fact specifically concerned with these. The findings thus demonstrated that an edutainment agenda is capable of not only entertaining visitors, but also educating visitors about difficult histories. It is suggested that this is further enhanced with the use of multiple methods within the interpretation design, including hands-on activities, innovative technologies, light and sound design, character actors, and exhibits.

By exploring the influences on the design of edutainment interpretation, the findings of this study contribute to dark tourism understanding as they contradict much of the literature that argues LDVAs are generally unconcerned with matters of historical accuracy and more often than not, trivialise history through myth-making. In addition, by exploring the influences on interpretation, this paper has contributed to the progress of this under-developed topic in dark tourism research. Subsequently, this paper contributes to the practical understanding of interpretation design within dark tourism, specifically edutainment interpretation at LDVAs. Finally, this paper argues greater attention in future research is required to continue the development of research and practical understanding of LDVAs and their edutainment interpretation agendas.

Acknowledgements

Thank you to the anonymous reviewers and Dr Rami Isaac for their helpful feedback and support. In addition, Prof. Anna Leask and Prof. Paul Barron provided valuable guidance and encouragement in the completion of this paper and formerly completed PhD. Also, I am grateful to the LDVAs and participants for their time given to this study.

Data availability

The data that support the findings of this study are openly available in Edinburgh Napier University Research Repository at http://researchrepository.napier.ac.uk/Output/2455103, 338.4791 Tourist Industry.

Disclosure statement

This paper draws on the author's PhD research (2019), which was partially funded by Edinburgh Napier University's 50th Anniversary Scholarship. Dr Brianna Wyatt is the author of this paper and supporting PhD research. Prof. Anna Leask served as the author's PhD Director of Studies. Prof. Paul Barron served as the author's PhD Supervisor. The authors declare that they have no known competing financial interests or personal relationships that could have appeared to influence the work reported in this paper.

ORCID

Brianna Wyatt http://orcid.org/0000-0002-2107-889X
Anna Leask http://orcid.org/0000-0003-3543-8064
Paul Barron http://orcid.org/0000-0002-6114-1187

References

Ashworth, G. J., & Page, S. (2011). Urban tourism research: Recent progress and current paradoxes. *Tourism Management*, *32*(1), 1–15. https://doi.org/10.1016/j.tourman.2010.02.002

Åstrøm, J. K. (2020). Why theming? Identifying the purposes of theming in tourism. *Journal of Quality Assurance in Hospitality & Tourism*, *21*(3), 245–266. https://doi.org/10.1080/1528008X.2019.1658147

Beaumont, T. (2014, November). Your 60 s guide to the Black Death. *BBC History*. http://www.historyextra.com/feature/your-60-second-guide-facts-black-death-how-when-why

Beck, L., & Cable, T. (2002). *Interpretation for the 21st century: Fifteen guiding principles for interpreting nature and culture* (2nd ed.). Sagamore Pub.

Bell, S., & Morse, S. (2013a). How people Use rich pictures to help them think and Act. *Systemic Practice and Action Research*, *26*(4), 331–348. https://doi.org/10.1007/s11213-012-9236-x

Bell, S., & Morse, S. (2013b). Rich pictures: A means to explore the sustainable mind? *Sustainable Development*, *21*(1), 30–47. https://doi.org/10.1002/sd.497

Berenbroek, L. (2013). *Translating Educational Children's Books: A translation and analysis of Terry Deary's Horrible Histories series*. [Utrecht University] http://dspace.library.uu.nl/handle/1874/282022

Bristow, R. S., & Jenkins, I. S. (2020). Geography of fear: Fright tourism in urban revitalization. *Journal of Policy Research in Tourism, Leisure and Events*, *12*(2), 262–275. https://doi.org/10.1080/19407963.2019.1631319

Brochu, L. (2003). *Interpretive planning: The 5-M model for successful planning*. InterpPress.

Bryon, J. (2012). Tour guides as storytellers: From selling to sharing. *Scandinavian Journal of Hospitality and Tourism*, *12*(1), 27–43. https://doi.org/10.1080/15022250.2012.656922

Carney, J. D. (1994). A historical theory of art criticism. *Journal of Aesthetic Education*, *28*(1), 13. https://doi.org/10.2307/3333153

Clarke, V., & Braun, V. (2017). Thematic analysis. *The Journal of Positive Psychology*, *12*(3), 297–298. https://doi.org/10.1080/17439760.2016.1262613

Dalton, D. (2015). *Dark tourism and crime*. Routledge.

Dunkley, R. (2017). A light in dark places? Analysing the impact of dark tourism experiences on everyday life. In G. Hooper & J. J. Lennon (Eds.), *Dark tourism: Practice and interpretation* (pp. 108–120). Routledge.

Dwyer, O., & Alderman, D. (2008). *Civil rights memorials and the geography of memory*. Center for American Places at Columbia College Chicago.

Foley, M., & Lennon, J. J. (1996). JFK and dark tourism: A fascination with assassination. *International Journal of Heritage Studies*, *2*(4), 198–211. https://doi.org/10.1080/13527259608722175

Fonseca, A. P., Seabra, C., & Silva, C. (2016). Dark tourism: Concepts, typologies and sites dark tourism: A troublesome concept. *Journal of Tourism Research and Hospitality*, *S2*(002), 1–6. http://doi.org/10.4172/2324-8807.S2-002

Gentry, G. (2007). Walking with the dead: The place of ghost walk tourism in Savannah, Georgia. *Southeastern Geographer*, *47*(2), 222–238. https://doi.org/10.1353/sgo.2007.0023

Goulding, C., & Domic, D. (2009). Heritage, identity and ideological manipulation: The case of Croatia. *Annals of Tourism Research*, *36*(1), 85–102. https://doi.org/10.1016/j.annals.2008.10.004

Grimwade, G., & Carter, B. (2000). Managing small heritage sites with interpretation and community involvement. *International Journal of Heritage Studies*, *6*(1), 33–48. https://doi.org/10.1080/135272500363724

Hannigan, J. (1998). *Fantasy city: Pleasure and profit in the postmodern metropolis*. Routledge.

Heidelberg, B. A. W. (2015). Managing ghosts: Exploring local government involvement in dark tourism. *Journal of Heritage Tourism*, *10*(1), 74–90. https://doi.org/10.1080/1743873X.2014.953538

Hertzman, E. (2006). *Visitors' evaluations of the historic content at Storyeum: An edutainment heritage tourist attraction*. [University of British Columbia]. https://open.library.ubc.ca/cIRcle/collections/ubctheses/831/items/1.0092637

Hertzman, E., Anderson, D., & Rowley, S. (2008). Edutainment heritage tourist attractions: A portrait of visitors' experiences at Storyeum. *Museum Management and Curatorship*, *23*(2), 155–175. https://doi.org/10.1080/09647770802012227

Heuermann, K., & Chhabra, D. (2014). The darker side of dark tourism: An authenticity perspective. *Tourism Analysis*, *19*(2), 213–225. https://doi.org/10.3727/108354214X13963557455766

Ho, J. K. K. (2015). An updated review on the conventional and unconventional rich picture building exercises (RPBEs) in P. B. Checkland's soft systems methodology. *American Research Thoughts*, *1*(7), 1516–1527.

Holloway, J. (2010). Legend tripping in spooky spaces: Ghost tourism and infrastructures of enhancement. *Environment and Planning D: Space and Society*, *28*(4), 618–637. https://doi.org/10.1068/d9909

Hooper, G. (2017). Introduction. In G. Hooper & J. J. Lennon (Eds.), *Dark tourism: Practice and interpretation* (pp. 1–11). Routledge.

Hovi, T. (2008). Tradition and history as building blocks for tourism: The middle ages as a modern tourism attraction. *Valahian Journal of Historical Studies*, *10*, 75–85.

Ioannides, D., Halkier, H., & Lew, A. (2014). Evolutionary economic geography and the economies of tourism destinations. *Tourism Geographies*, *16*(4), 535–539. https://doi.org/10.1080/14616688.2014.947315

Isaac, R. K., & Çakmak, E. (2014). Understanding visitor's motivation at sites of death and disaster: The case of former transit camp Westerbork, the Netherlands. *Current Issues in Tourism*, *17*(2), 164–179. https://doi.org/10.1080/13683500.2013.776021

Ivanova, P., & Light, D. (2018). 'It's not that we like death or anything': Exploring the motivations and experiences of visitors to a lighter dark tourism attraction. *Journal of Heritage Tourism*, *13*(4), 356–369. https://doi.org/10.1080/1743873X.2017.1371181

Jamal, T., & Lelo, L. (2011). Exploring the conceptual and analytical framing of dark tourism: From darkness to intentionality. In R. Sharpley & P. Stone (Eds.), *Tourist experience: Contemporary perspectives* (pp. 29–42). Routledge.

Jones, S. (2007). *Sharing Our Stories: Guidelines for Heritage Interpretation.*

Kavanagh, G. (1996). *Making histories in museums.* Leicester University Press.

Kelly, J. (2006). *The great mortality: An intimate history of the Black death, the most devastating plague of all time* (2nd ed). Harper Perennial.

Kidd, J. (2011). Performing the knowing archive: Heritage performance and authenticity. *International Journal of Heritage Studies*, *17*(1), 22–35. https://doi.org/10.1080/13527258.2011.524003

Kim, S., & Butler, G. (2015). Local community perspectives towards dark tourism development: The case of Snowtown, South Australia. *Journal of Tourism and Cultural Change*, *13*(1), 78–89. https://doi.org/10.1080/14766825.2014.918621

King, M. (1991). The theme park experience: What museums Can learn from Mickey Mouse. *The Futurist*, *25*(6), 24–31.

Knudsen, B., Cable, T., & Beck, L. (1995). *Interpretation of cultural and natural resources.* Venture Pub.

Korstanje, M. (2018). Research methods in dark tourism fields. In M. Korstanje & B. George (Eds.), *Virtual traumascapes and exploring the roots of dark tourism* (pp. 84–98). IGI Global.

Leask, A. (2016). Visitor attraction management: A critical review of research 2009–2014. *Tourism Management*, *57*, 334–361. https://doi.org/10.1016/j.tourman.2016.06.015

Light, D. (2017). Progress in dark tourism and thanatourism research: An uneasy relationship with heritage tourism. *Tourism Management*, *61*, 275–301. https://doi.org/10.1016/j.tourman.2017.01.011

Magee, R., & Gilmore, A. (2015). Heritage site management: From dark tourism to transformative service experience. *The Service Industries Journal*, *35*(15–16), 898–917. https://doi.org/10.1080/02642069.2015.1090980

McEvoy, E. (2016). *Gothic tourism.* Palgrave Macmillan.

Oren, G., & Shani, A. (2012). The Yad Vashem Holocaust museum: Educational dark tourism in a futuristic form. *Journal of Heritage Tourism*, *7*(3), 255–270. https://doi.org/10.1080/1743873X.2012.701630

Platt, C. (2014). *King death: The Black death and its aftermath in late-Medieval England.* Routledge.

Potter, A. (2016). She goes into character as the lady of the house: Tour guides, performance, and the Southern plantation. *Journal of Heritage Tourism*, *11*(3), 250–261. https://doi.org/10.1080/1743873X.2015.1100626

Powell, R., & Iankova, K. (2016). Dark London: Dimensions and characteristics of dark tourism supply in the UK capital. *Anatolia: An International Journal of Tourism and Hospitality Research*, *27*(3), 339–351. https://doi.org/10.1080/13032917.2016.1191764

Quinn, B., & Ryan, T. (2016). Tour guides and the mediation of difficult memories: The case of Dublin Castle, Ireland. *Current Issues in Tourism*, *19*(4), 322–337. https://doi.org/10.1080/13683500.2014.1001727

Rátz, T. (2006). Interpretation in the house of terror, Budapest. In M. Smith & M. Robinson (Eds.), *Cultural tourism in a changing world: Politics, participation and (Re)presentation* (pp. 244–256). Channel View Publications.

Roberts, T. (2014). Interpretation design: An integrative, interdisciplinary practice. *Museum & Society*, *12*(3), 191–209.

Roberts, T. (2015). Factors affecting the role of designers in interpretation projects. *Museum Management and Curatorship*, *30*(5), 379–393. https://doi.org/10.1080/09647775.2015.1055582

Rodriguez-Garcia, B. (2012). Management issues in dark tourism attractions: The case of ghost tours in Edinburgh and Toledo. *Journal of Unconventional Parks, Tourism & Recreation Research*, *4*(1), 14–19.

Ron, A. S., & Timothy, D. J. (2013). The land of milk and honey: Biblical foods, heritage and Holy Land tourism. *Journal of Heritage Tourism*, *8*(2–3), 234–247. https://doi.org/10.1080/1743873X.2013.767817

Santonen, T., & Faber, E. (2015). Towards a comprehensive framework to analyse edutainment applications. *ISPIM Conference Proceedings*, *358*(June), 1–11.

Scanlon, M. (2011). History beyond the academy: Humor and horror in children's history books. *New Review of Children's Literature and Librarianship*, *16*(2), 69–91. https://doi.org/10.1080/13614541.2010.540197

Shani, A., & Logan, R. (2010). Walt Disney's world of entertainment attractions. In R. Butler & R. Russell (Eds.), *Giants of tourism* (pp. 155–169). CABI Publishing.

Silverman, H. (2011). Contested cultural heritage: A selective historiography. In H. Silverman (Ed.), *Contested cultural heritage: Religion, nationalism, erasure and exclusion in a global world* (pp. 1–49). Springer.

Skov, M. B., & Andersen, P. B. (2001). Designing Interactive Narratives General Terms. *Proceedings of the 1st International Conference on Computational Semiotics in Games and New Media*, 59–66.

Smith, M. (2016). *Issues in cultural tourism studies* (3rd ed). Routledge.

Spaul, M., & Wilbert, C. (2017). Guilty landscapes and the selective reconstruction of the past: Dedham Vale and the murder in the Red Barn. In G. Hooper & J. J. Lennon (Eds.), *Dark tourism: Practice and interpretation* (pp. 83–93). Routledge.

Stone, P. (2006). A dark tourism spectrum: Towards a typology of death and macabre related tourist sites, attractions and exhibitions. *TOURISM: An Interdisciplinary International Journal, 54*(2), 145–160.

Stone, P. (2009a). Dark tourism: Morality and new moral spaces. In R. Sharpley & P. Stone (Eds.), *The darker side of travel: The theory and practice of dark tourism* (pp. 56–72). Channel View Publications.

Stone, P. (2009b). "It's a bloody guide": Fun, fear and a lighter side of dark tourism at the dungeon visitor attractions. In R. Sharpley & P. Stone (Eds.), *The darker side of travel: The theory and practice of dark tourism* (pp. 167–185). Channel View Publications.

Stone, P. (2013). Dark tourism scholarship: A critical review. *International Journal of Culture, Tourism and Hospitality Research, 7*(3), 307–318. https://doi.org/10.1108/IJCTHR-06-2013-0039

Stone, P. (2018). Dark tourism in an age of 'spectacular death'. In P. R. Stone, R. Hartmann, T. Seaton, R. Sharpley, & L. White (Eds.), *The Palgrave handbook of dark tourism studies* (pp.189–210). Palgrave Macmillan.

Tarlow, P. (2005). Dark tourism: The appealing dark side of tourism and more. In M. Novelli (Ed.), *Niche tourism: Contemporary issues, trends and cases* (pp. 47–58). Elsevier.

Watson, S. (2018). The legacy of communism: Difficult histories, emotions and contested narratives. *International Journal of Heritage Studies, 24*(7), 781–794. https://doi.org/10.1080/13527258.2017.1378913

Welch, M. (2016). Political imprisonment and the sanctity of death: Performing heritage in 'troubled' Ireland. *International Journal of Heritage Studies, 22*(9), 664–678. https://doi.org/10.1080/13527258.2016.1184702

Wight, A. C. (2006). Philosophical and methodological praxes in dark tourism: Controversy, contention and the evolving paradigm. *Journal of Vacation Marketing, 12*(2), 119–129. https://doi.org/10.1177/1356766706062151

Wight, A. C., & Lennon, J. J. (2007). Selective interpretation and eclectic human heritage in Lithuania. *Tourism Management, 28*(2), 519–529. https://doi.org/10.1016/j.tourman.2006.03.006

Woodward, M. (2009). *Overlapping dialogues: The role of interpretation design in communicating Australia's natural and cultural heritage.* [Curtin University of Technology] https://researchoutput.csu.edu.au/en/publications/overlapping-dialogues-the-role-of-interpretation-design-in-commun-3

Wyatt, B. (2019). *Influences on interpretation: A critical evaluation of the influences on the design and management of interpretation at lighter dark visitor attractions* [Edinburgh Napier University]. https://www.napier.ac.uk/~/media/worktribe/output-2455103/influences-on-interpretation-a-critical-evaluation-of-the-influences-on-the-design-and.pdf

Uncomfortable and worthy: the role of students' field trips to dark tourism sites in higher education

Ilze Grinfelde ⓘ and Linda Veliverronena

ABSTRACT

This article studies the importance of the educational experience of students in dark tourism sites, with a particular focus on post-visit effects after their exposure to dark sites – inner-directed and behavioural activities. Previous studies on the topic are limited, controversial, and reveal that field trips as a compulsory activity do not necessarily induce post-visit effects, but at the same time can provoke greater interest in certain topics, such as research or repeated visits. The study continues the strand of research of post-visit effects after exposure to dark tourism sites of the specific segment – university undergraduates. The study is based on longitudinal data collected after field trips to dark tourism sites in Latvia (2014–2019). In total, 119 Latvian and exchange students participated in field trips during the Dark tourism study course. Students reported about the post-visit effects by using questionnaires based on dark tourism theoretical frameworks. The programme included such dark tourism sites as the KGB Museum, Ghetto Museum, and memorials. The results indicate that students appreciate the educational value of the dark tourism sites and better-designed sites induce more post-visit effects. However, educational gains must be strengthened by stimulating professional knowledge and the will to research independently.

Introduction

According to Seaton (1996), dark tourism experiences are sublime, numinous, even mystical, and extraordinary. Dark tourism can provide an emotional and cognitive space that leaves a trace with the physical one. There is a high potential for a dark place to have a strong impact on visitors – their relation and interaction with space (Martini & Buda, 2018; Yan et al., 2016). Cooke (2012) has noted that visits to the dark tourism sites provoke complex reactions in people, although these trips can be undertaken for reasons that might not follow dark motivations, such as natural fascination by death or education (Cooke, 2012; Sather-Wagstaff, 2011).

Generally, dark tourism research recognizes learning as one of the motives for visiting dark sites. Beneficial post-visit outcomes from the perspective of the visitors are education and increased knowledge, as well as the transforming power of individuals' consciousness (Biran et al., 2011; Cohen, 2011; Lennon & Foley, 1999; Stone, 2012; Walter, 2009). From a rational perspective, educational visits to dark tourism sites are an often adopted alternative and they significantly differ from general leisure visits as their aim is to meet certain learning objectives (Sharpley & Baldwin, 2009).

In formal education, a field trip to a dark tourism site is considered as a method in the integrated, competence raising picture for the visitors. For example, visits are a tool to broaden understanding of the history, literature, and art inspired by certain events. The field trips also serve as a pedagogical framework to explore the spiritual issues and reactions of different faiths and stimulate historical empathy. Verification of the history through the exploration of artefacts and authentic sites can be also an entertaining experience through simulation (Morrison, 2019; Sharpley & Baldwin, 2009; Walter, 2009).

The aim of this article is to study the importance of the educational experience of students in the dark tourism sites, with a particular focus on post-visit effects after their exposure to dark sites – inner-directed and others-directed behavioural activities. With inner-directed post-visit effects, we understand behavioural activities guided by internal values, e.g. such as reading and researching more about the topic, rethinking students' personal values, reconsidering individual and collective responsibility, and rethinking the past and future. With others-directed behavioural effects, we understand activities that involve interaction with other people, such as talking with peers, telling about the site to friends or family, planning to revisit a site with friends, etc.

Although the strong focus on the integration of dark tourism experiences in formal education activities is lacking in academic studies, there are studies examining the dualism of education and tourism and the interrelationship between these two purposes (Cooper, 2006; Lennon & Foley, 2000; Stone, 2006; Yoshida et al., 2016). After reviewing previous studies, Isfrailova and Khoo-Lattimore (2018) conclude that despite the fact that the potential of dark tourism sites as an educational tool was validated, the educational element has never been the main focus. In particular, studies of the impact of dark tourism sites on university students who visit them as part of their field trips (within or outside specific dark tourism-related courses) are more than limited. Therefore, in this study, we focus on exploring the educational significance of the dark sites and young adult visitors' post-visit experiences.

The turbulent history of Latvia in the twentieth century has generated a wide range of sites representing the whole spectrum of dark tourism. However, previous studies have been mostly descriptive and focused on collective memory, while the political and tourism perspectives are rather rarely investigated, specifically in the context of Baltic countries (Isaac & Budryte-Ausiejiene, 2015; Lankauskas, 2006). Overall, the educational role of dark tourism is not sufficiently discussed. On a Baltic scale, there are no previous studies exploring the educational role of dark tourism sites although visits to dark tourism sites have been a part of youth education at different levels – starting from primary school.

The study is based on longitudinal data collected after study excursions to several dark tourism sites in Latvia from 2014 to 2019 held as part of a dark tourism study course for bachelor students. During study excursions, students attended three to four dark tourism sites and afterwards reported in a survey their emotional reactions and effect on their further activities (e.g. searching for further information, sharing and discussing their experience, and recommending others to visit a site). Research data were analysed with descriptive statistical data analysis methods. The majority of the visited sites are related to events of World War I, World War II, and the post-war period and involve both Nazi and Communist regimes. Heritage sites associated with Nazi and Communist regimes are underresearched in the Baltic States in a dark tourism context because of shortage of information, their sensitivity and high emotional load, and the events' temporal proximity (Grinfelde & Veliverronena, 2018). Additionally, students' feedback about the course performance was analysed by using the method of qualitative thematic analysis and we focus on the content describing the role of the field trip in the study course.

The paper begins with a literature review on dark tourism in formal education and a study trip as a learning tool. It is followed by a contextualization of the study and description of data collection and analysis methods. The findings are followed by the conclusive discussion regarding the study results.

Dark tourism in the context of education

There are a number of reasons why people visit dark tourism sites – curiosity, survivor guilt, remembrance, nostalgia, empathy, horror, etc. Gaining new knowledge and understanding are among them too (Kang, 2010). Specifically, heritage objects provide various informal and formal education contexts. For example, Kang (2010) points out that the *raison d'etre* of certain sites (e.g. memorials of past events) is purely educational. According to Chang (2017), visitors' perception of dark tourism affects their geopolitical knowledge, e.g. stimulates a sense of territory. Also, after the visit of a dark site, visitors have a desire to understand the past history of conflicts.

Despite the educational potential of dark heritage sites, not all dark tourism places provide new knowledge and understanding. For example, on the basis of previous studies, Korstanje and George (2017) identify three broad types of dark tourism sites: (1) profit-oriented; (2) interaction-oriented; (3) heritage-oriented. Heritage-oriented sites emphasize the importance of experience and connect visitors to human sufferings and as such may have greater educational potential.

Stone (2006) argues dark tourism products are multifarious in design and characteristically varied and offers the concept of a dark tourism spectrum – darkest, darker, dark, light, lighter, and lightest products. The products in the darkest category are highly emotional, have higher political and ideological influences, and are education-oriented and history-centric while those in the lightest category are entertainment-oriented and heritage-centric and less political and ideological (Stone, 2006). From this point of view, the darkest tourist sites would be the most suitable for education activities. However, according to Shackley (2001) (as cited in Stone, 2006), they are tricky to interpret. Stone (2006, p. 150) also refers to other sources stating that 'the heritage sector in general is an inappropriate and even immoral vehicle for the presentation of death'. Similarly, commenting on dark exhibitions, Stone (2006) indicates that educative elements of the dark exhibitions are undoubted and yet commercialization has perhaps tainted the original exhibit objectives and reminds us that many battlefields have become romanticized and trivialized through a fun approach. Overall, commercialization or commodification can at times accompany the educational aspects of a visit. This is the challenging aspect in organizing study trips to dark tourism sites.

An indirect indicator of students' and pupils' interest in the darkest and education-oriented heritage sites is identified in previous studies examining students' motivation and expectations in formal education. For instance, Isfrailova and Khoo-Lattimore (2018) have studied how visiting dark tourism sites impacted school children, and the findings suggest children's visit to a thanatological site fills their knowledge gap and motivates them to study history. The situation differs in regard to young adults studying in universities. A recent study (Grinfelde & Veliverronena, 2019) reveals a majority of bachelor tourism students in higher education institutions in Latvia assess study quality and meaningfulness not only in terms of new knowledge and skills but even more in terms of entertainment and personal interests. Frequently, students are less concerned about the actual content of the study courses but rather focused on the organizational issues and packaging of study content (e.g. an entertaining and interesting style of teaching). This also indirectly indicates the dilemma faced by teachers: how to select dark tourism sites for compulsory field trips to increase students' knowledge on the topic and at the same time keep their interest alive. Kang et al. (2012, p. 261) have examined visitor experiences, including during a compulsory school field trip, in the Jeju April 3rd Peace Park in Korea and argue that 'a compulsory field trip program may not stimulate a visitor's interest in the incident, and thus cannot necessarily generate effective visitor learning and emotional experiences'. Kang (2010) also indicates the importance of visitor engagement in relation to gains from visiting a dark tourism site. He has also found out that older visitors with higher education and those residing locally are more likely to show high involvement and consequently benefit more from a visit.

A study trips as an effective learning tool

Recent trends in tourism research and hospitality education cover the fields of leadership and human capital development. To develop these qualities among students, teaching methods with an emphasis on active and experiential learning, online education, diversity education, internationalization, and industry experience are used (Kim & Jeong, 2018). As Fisher et al. (2017) have stated, not all industry-relevant skills can be achieved through academic curricula in a traditional classroom setting. Tal and Morag (2009) described formal study excursions as student experiences outside of the ordinary classroom at interactive locations with a previously designed program with educational purposes. Authors emphasize that field trips provide first-hand experience, stimulate interest and motivation in a specific field, consolidate theory and practice, develop and strengthen observation and perception skills, and promote personal development. However, it is important to note that the role of the teacher as a field trip leader and facilitator is different as in a classroom setting because an instructor is involved and participates in the preparation and field trip activities, including post-visit reflections. Educational visits or field trips are considered as an experiential learning tool. It is a student-centred approach which is embedded in course curricula in higher education and evaluated from the point of view of students as worthy, stimulating important connections to course content through the provision of an opportunity to explore the topics of study in real-life settings (Djonko-Moore & Joseph, 2016; Pattachini, 2018).

Besides rationalized needs of curriculum, study excursion components can stimulate learning which according to Mezirow (1978) transforms problematic frames of reference to make students more inclusive, reflective, open, and emotionally able to change. The core meaning of such a learning process called transformative learning is in constructing and assigning new and adjusted interpretations of the meaning of an experience in order to guide future action. Kirillova et al. (2016) has stated that post-trip anxiety motivates tourists to resolve personal existential dilemmas and to initiate meaningful life changes. A few studies connect to recent theoretical conceptualizations of transformative learning as it connects to the affective, emotional, and embodied aspects. The above-mentioned is also associated with discomfort requesting bravery and courage from both students and educators (Walker & Manjamba, 2020). There can be an obvious link between a visit to a dark tourism site and transformative learning, as has been stated in Walker and Manjamba (2020, p. 6.)–'the shame and guilt we experience upon witnessing another's pain, being confronted by our own egotism and privilege, are common disorienting dilemmas in travel'. The role of emotions might suggest some advantages of field trips for learning in formal education because during a travel experience, visitors may have so-called peak experiences when their senses and cognitive abilities are exacerbated (Biran et al., 2011).

In the context of dark sites and post-effects of educational visits, it is worth remembering that emotions have an influence on the cognitive processes of humans, including perception, learning, memory, reasoning, and problem solving and attention (Timm Knudsen, 2011). Wang and Carlson (2011) have concluded after reviewing previous studies that the emotions of students have been neglected in most field trip evaluations and it is rarely considered what students are 'feeling' during a field trip. In the field of educational psychology, a lot of studies (Mortari, 2015; Tyng et al., 2017; Wang & Carlson, 2011) support the ability of emotions to change people's thoughts, actions, and physiological responses.

Context of the study and research methods

In recent years, Dark tourism as an academic field of study has appeared in universities' study programmes with an aim to cover different perspectives – scientific, entrepreneurial, historic, and also personal development. The study course 'Dark tourism and visitor motivation' has been included in the bachelor's level curricula of Tourism and hospitality management studies of Vidzeme

University of Applied Sciences (Latvia) since 2013. This course is designed to provide students with knowledge about the foundations of dark tourism, diverse forms, and management of dark tourism sites.

The main study methods used in the implementation of the course are lectures, discussions, workshops, presentations, individual and group assignments, 1 local dark site visit, and 1 field trip. Participants of the course are mainly 3rd year Tourism and hospitality management students but also exchange students take part and it is an elective course for the students from other study directions. The small group size (25–45 students) allows for more flexibility in the application of specific study methods and a more individual approach as the course content can be uncomfortable and emotionally demanding.

Study visits to dark tourism sites take place as a mandatory part in the Dark Tourism study course in 2014–2019. The study excursion is usually planned by the course tutor and implemented in the middle of the course, when students already have had insights into theoretical aspects of dark tourism and have been acquainted with the specifics of emotions of dark tourism site visitors. The duration of the excursion is 1 day, it includes visits to three to four dark tourism sites to balance the intensity of emotions and leaves enough time for individual on-site exploration and peer discussions. The choice of the sites is based on the principle that they should represent different aspects of the dark tourism spectrum and variety of tourism products. The examples of the sites are memorials (The Brethren Cemetery, Rumbula Jews memorial, Salaspils concentration camp memorial), cemeteries (Raiņa cemetery, The Grand Cemetery of Riga), museums (Lipke memorial, Riga Ghetto museum, and former ghetto territory, KGB museum, War museum). During 2014–2019 altogether nine different sites were visited, some of them were visited each year, while others only once.

Before the trip students receive information about the trip program, information about the sites, suggestions for further reading or movies to get more acquainted with the context of the historical events, involved personalities, etc. One week after the trip, the feedback session (four academic hours) is organized. The students receive questionnaire forms which are adapted from Best (2007) and Fokkinga and Desmet's (2013) studies into study experience evaluation. The survey contains two sets of questions. In the first set, respondents evaluate their emotions. The second part of the questionnaire investigates if the visit has provoked some post-visit effects, firstly, inner directed such as reading and researching more about the topic, rethinking their personal values, rethinking individual and collective responsibility, rethinking the past and future, and, secondly, talking with peers, telling about the site to friends or family, planning to revisit the site, etc. In the questionnaire, a 5-point (0–4) Likert scale is used. The form also includes a commentary section where students can write about their feelings or what they think by asking open-ended questions. We use these comments to illustrate quantitative data.

The obvious limitation of the study is related to the sample size. Each year, the number of students has varied, as has the frequency of visitation of different sites. As a result, the number of respondents for each site varies between 12 and 103. For this reason, we have performed data analysis only for sites visited by at least 30 students (see Table 1), as in practical applications, 'the distribution of sample mean may be assumed to be normal distributions if the sampling size is larger than 30' (Chang et al., 2006, p. 31). Despite considering this as a research limitation, we would also like to emphasize that, in given conditions, we have included in the sample all the students attending the Dark tourism study course.

Descriptive statistical analyses for mean (further in text AVG), median, and standard deviation values were performed to describe students' post-visit effects in these six dark tourism sites. The positions of the sites in the dark tourism spectrum were taken into account in the data analysis.

After completing the study course, students receive standardized questionnaires exploring their opinion about the study course as part of the university quality system. It consists of closed- and open-ended questions focusing on course quality, student satisfaction, personal gains, and the instructor performance. Filling out this questionnaire is voluntary. The total number of questionnaires filled in 2014–2019 is 70. The answers to the open-ended questions about the weaknesses and

Table 1. Description of the visited dark tourism sites and their position in the dark tourism spectrum according to Stone (2006).

Site	Visits	Description of the site and context of the visit	Place in dark tourism spectrum
State security agency (KGB) building, called Corner house	103	During the Soviet occupation, the state security agency (KGB) imprisoned, tortured, killed its victims in the building which nowadays is a part of the Occupation museum although the site itself is left in its authentic condition. One of the darkest tourism sites in Latvia. Students had a guided tour of its cellars. Besides that, students had access to an informative exposition with elements of video witnessing.	Darkest
Rumbula Jews memorial	75	Rumbula is one of the largest holocaust sites in Europe. The design of the memorial represents the Rumbula tragedy of a pre-planned massacre in winter of 1941 in the outskirts of the capital of Latvia, Riga. The memorial is one of the darkest tourism sites in Latvia. At the site of the massacre, places of mass graves are marked and memorials devoted to past events. Students were introduced to the site by the teacher.	Darkest
Riga Ghetto museum	66	The Riga Ghetto Museum is located in the historic area bordering the former ghetto. It features more than 70,000 names of Holocaust victims and a photo exhibition focusing on anti-Semitic propaganda, the Holocaust in Latvia, the resistance movement, and those who provided refuge for Jewish people. Students were introduced to the site by the teacher.	Dark
Brethren Cemetery	55	The Brethren cemetery is the memorial ensemble dedicated to the soldiers who fell during World War I and the Latvian Freedom Fights (1915–1920). Around 3000 soldiers, including 300 unknown ones, are buried in the Cemetery ground. Students were introduced to the site by the teacher.	Light
The Salaspils Memorial Ensemble	48	The Salaspils Memorial Ensemble was an extended police prison and labour correctional camp of National Socialistic Germany (1941–1945). It was built by deported Jews from Germany, Austria, and Czechoslovakia. The camp imprisoned Latvian, Lithuanian, and Estonian civilians and military personnel, as well as transit prisoners, including small children from Latgale, Belarus, and Russia. Over 20,000 individuals found themselves there, and at least 2000–3000 people died in the camp. Students were introduced to the site by the teacher.	Dark
Lipke memorial	41	The memorial is dedicated to Žanis Lipke and his life, for during the Second World War, he saved from death more than 50 Jews by hiding them in a bunker specially designed under his shed. The memorial building was designed so that it would represent the living conditions of people hiding there as closely as possible. Students had a guided tour.	Light

Source: Site descriptions are based on the information available on objects' websites and official tourism sites.

strengths of this study course are used as additional data to examine the role of the field trip in the dark tourism study course. The texts are carefully read and the content regarding students' comments is analysed regarding following themes – post-visit effects (e.g. reading, thinking, plans to revisit, etc. and role of field trip in a context of study course, etc.).

Findings

This section is focused on the post-visit effects experienced by the students after visiting dark tourism sites. Quantitative data from the questionnaire are enriched by the open-end questions. The quotes illustrate the significance of the post-visit effects provoked during the field trip. For example, the quote below has been said about the Rumbula memorial:

> Very impressive, sad place. I was thinking about it for some week …

As stated before, descriptive statistics for the mean, median, and standard deviation values of post-visit effects identified by students (see Table 2) have been analysed. Respondents evaluated their

Table 2. Statistical average values rates and median of post-visit effects identified by students.

	Research about topic	Re-evaluate	Rethink responsibility	Rethink past	Rethink future	Retalk with course mates	Tell to others	Revisit	Recommend visit	Visit other objects
KGB museum[a]	2.63	2.90	2.92	3.72	2.53	3.11	3.56	2.57	3.54	3.03
KGB museum[b]	3.00	3.00	3.00	4.00	3.00	4.00	4.00	3.00	4.00	3.00
KGB museum[c]	1.35	1.16	1.15	0.58	1.31	1.13	0.86	1.53	0.91	1.14
Ghetto museum[a]	1.85	2.35	2.35	3.29	2.03	2.47	3.03	1.53	2.56	2.20
Ghetto museum[b]	2.00	2.50	2.50	4.00	2.00	3.00	3.00	1.00	3.00	2.00
Ghetto museum[c]	1.32	1.26	1.22	0.96	1.45	1.22	1.14	1.43	1.34	1.28
Brethren cemetery[a]	1.54	1.80	1.95	3.11	1.66	2.15	2.64	1.73	2.62	2.09
Brethren cemetery[b]	1.00	2.00	2.00	4.00	1.00	2.00	3.00	2.00	3.00	2.00
Brethren cemetery[c]	1.37	1.39	1.27	1.21	1.43	1.35	1.41	1.45	1.50	1.42
Rumbula Jews memorial[a]	2.04	2.49	2.72	3.44	2.37	2.77	3.08	1.47	2.65	2.31
Rumbula Jews memorial[b]	2.00	3.00	3.00	4.00	2.50	3.00	3.00	1.00	3.00	2.00
Rumbula Jews memorial[c]	1.32	1.33	1.31	0.92	1.35	1.32	1.10	1.38	1.22	1.38
The Salaspils Memorial Ensemble[a]	1.87	2.22	2.34	3.11	2.23	2.60	3.11	2.24	3.45	2.32
The Salaspils Memorial Ensemble[b]	2	2	2	3	2	3	3	2	3	2
The Salaspils Memorial Ensemble[c]	1.28	1.19	1.26	1.05	1.29	1.14	1.01	1.21	1.15	1.29
Lipke Memorial[a]	2.21	2.69	2.49	3.23	2.10	2.44	3.18	2.10	3.05	2.59
Lipke Memorial[b]	2	3	3	4	2	3	4	2	3	3
Lipke Memorial[c]	1.30	1.32	1.23	1.06	1.33	1.27	1.12	1.35	1.12	1.35

[a]Mean (AVG).
[b]Median.
[c]Standard deviation.

experiences by using the Likert scale where the answer '0' meant the visit made no impact at all and '4' indicated a significant impact on the respondent's behaviour. Overall, descriptive statistics present that visits to the sites have motivated students to reconsider past events which are typical for every site as this post-visit effect shows the highest average value rates.

The most intense effects students have experienced have been after visiting the KGB Museum – one of the darkest tourism sites in Latvia and a very authentic one (valid responses from 103 students, consequently it was the most frequently visited dark tourism site). The characteristics of the KGB Museum probably explain why this site has left the most significant effects on all types of students' post-visit behaviour:

> The people's stories made me cry sometimes. The field trip made me think about the real values in life. What is really important for us and how short was life for these people, who stayed here.

Likewise, many other quotes written by students unveil strong emotions experienced in the KGB Museum, new knowledge, and discoveries about the past events. According to the answers given by the students to the open questions, emotional overload has somewhat hindered the evaluation of the museum as a tourism product and the focus on it from a professional perspective, which is a significant aspect of the field trip.

The KGB Museum has induced such effects as rethinking about the past and future, revising one's own values, sharing the information about the site and experience, recommendation to revisit or visit similar objects, while experience in other sites mostly indicates inner-directed processes. Answers about post-visit effects in the KGB Museum display a higher median and lower mean, which indicates polarized opinions among students. For example, the majority of students have discussed their KGB on-site experience with other classmates, while few have been unwilling to do so. Probably these polarized opinions could be explained by uncomfortable feelings as quite a few students refer in their open remarks to uneasiness, confusion, inability to read their emotions, or understand what they have experienced at the KGB Museum. The quote below illustrates such remarks:

> Day started with KGB and we all got a big SHOCK and a lot of emotions, most of us took a while to understand what we actually feel and think.

Another site that has stimulated both self-reflection and others-directed behavioural activities is the Rumbula Jews Memorial. The number of valid responses in this case is 75. The Rumbula Jews Memorial is located in exactly the same place where the mass killings took place; however, the place itself does not display authentic elements anymore and is constructed around a recently-built memorial. The transformation was noted by one student: 'Powerful site, but it does not transmit the same feelings as the other site (meant KGB) because the original buildings are not there.' Still, the Rumbula Jews Memorial is another site which students have perceived rather emotionally, just like the KGB Museum. Results show high values in all the aspects. However, students' intent to revisit the place is below the average (AVG 1.47):

> I went home with 'a brick on my heart' thinking about terror and absurdity that fellow humans can inflict upon others. Actually it scares me, thinking about my future children, that something like this could happen again.

The site does not offer a guided tour by the professional guide as in the KGB Museum. The lack of tour guide service and authenticity might limit students' ability to relate themselves to the tragic past and to experience the extent of painful on-site emotions, while surprise and sadness provoked students to share this experience with others and recommend site visits to friends and relatives.

If students referred to the KGB and Rumbula Jews Memorial museums rather emotionally, then our observations of students' on-site behaviour and their comments in the questionnaires lead us to think very differently about the Riga Ghetto Museum. It was perceived mostly as a tourism object rather than a dark tourism site and the place was not attractive enough as a tourism site for a contemporary youngster:

> I was not excited about this place. They need to make this more attractive for visitors.' or 'It should be bigger with more interactive things.

Students' perception could be explained by the nature of the Ghetto Museum which is located in the very centre of the capital city, on the edge of the historical ghetto territory in Riga. Despite the fact the museum exhibits authentic elements, the overall authenticity of the site is low, and students do not experience intense emotions typical for dark tourism sites. To provide different dark tourism product interpretations during field trips, the group did not use a tour guide service in this museum, and therefore students had to introduce the museum exhibition themselves by reading and watching visuals.

The fourth most visited site we included in our analysis is the Brethren Cemetery, a classical memorial dedicated to WWI soldiers and from this point different from other sites that are all about victims of oppressive regimes. The Riga Brethren Cemetery is located on the lightest end of the dark tourism spectrum and has generated less significant post-visit effects than other sites according to the statistical mean. The median value displays very few students have been interested to do more extensive research on this topic (AVG 1.54) or revisit the site. However, the site has motivated students to rethink the past (AVG 3.11) and this variable displays median '4'.

The Salaspils Memorial Ensemble (valid responses $n = 48$) is established in the former site of a Nazi regime camp (1941–1945). The memorial consists of several impressive sculptures and buildings representing Soviet Brutalist architecture, which in combination with a location in a lively and detailed forest can give the visitor an impressive experience from different viewpoints. To some extent, we can say that the results are similar to the Ghetto Museum. However, it is worth emphasizing that this is one of the sites which motivated students not only to rethink the past (AVG 3.11) but to recommend a site visit to others (AVG 3.45, which is the second highest score for this post-visit effect).

The Lipke Memorial (valid responses $n = 41$) is designed with the purpose of a modern memorial, well-shaped aesthetically and keeping in mind the perspective of the tourism product. These aspects potentially play a role in the post-visit effects, as we see – the statistical means for most effects are high (AVG >3 in effects rethink the past, tell to others, recommend visit to others). However, we have to take into account that medians indicate polarized answers with regard to some effects, such as telling to others and rethinking the past, which means that some students were active while some were indifferent.

We can conclude that the visits to the dark tourism sites representing the darkest part of the dark tourism spectrum, in general, induce more post-visit effects than those from the lighter part. Answers are more homogenic in regard to darkest visited sites (lower standard deviation rates); however – all standard deviation rates were below 1.5. A lower mean was shown for rethinking the future; however, in the majority of cases, the values were above the average. In comparison with other effects, the lowest values were shown with regard to the will to do more research on the topic. However, from a pedagogical point of view, this is the effect that teachers would like to stimulate more. Sites representing various parts of the dark tourism spectrum (Lipke Memorial, KGB Museum, The Salaspils Memorial Ensemble) have stimulated students' will to recommend to other people a visit to particular sites. Also, respondents probably would revisit the sites mentioned above in the future. Considering descriptive statistics according to the post-visit effects, it is visible that answers about some activities have been more polarized (e.g. rethinking the past) than in the case of others (e.g. visiting other objects).

Additionally, students' feedback ($n = 70$) about the course performance was analysed. Textual comments were written in half of the questionnaires. The course performance assessment from students was also helpful in understanding the role of the field trip as a specific method within a course.

In the majority of comments, students mentioned the field trip and reflected that the study trip plays an essential role in a dark tourism course:

> Field trip was exceptional because I myself would not go to places like this. Therefore, it was a great opportunity to see and investigate something more.

In the comments, respondents indicated that the field trip helped to enrich their understanding of the topic and to remember the content, to identify the diversity of dark tourism, and to establish a balance between theory and practice. For many, the trip was an amazing chance from an individual (especially emotional) perspective, an eye-opener with regard to dark tourism site visits.

The comment section in the course performance evaluation questionnaires also revealed a few post-visit effects, such as motivation to investigate history, increased understanding of tourist behaviour, and reconsideration of values (e.g. patriotic feelings). However, the questionnaire was not designed to examine post-visit effects and the comments did not provide a full picture of effects.

Conclusion

In this study, we focused on exploring the role of the educational experience gained during the field trip in the context of a dark tourism study course with a particular focus on the post-visit effects after students' exposure to dark sites. The findings are in line with previous studies (Djonko-Moore & Joseph, 2016; Pattachini, 2018) and confirm the significance of field trips as students report professional benefits, such as increased knowledge of the topic, ability to link theory and practice, and extended understanding of different tourism products.

The visits to the darkest sites have been obviously uneasy experiences for students, but still these are the ones leaving the most significant post-visit effects. Thus, the visit to the KGB Museum motivates students to do post-visit research more than any other of the sites. The conclusions of our study offer support from a student's perspective to previous studies noting the darkest tourism sites are perceived to have higher educational value (e.g. Stone, 2006). Results also show that not all dark tourism sites included in the excursion program have generated effective visitor learning and emotional experiences as previously stated by Kang et al. (2012, p. 261) – instead, lighter dark tourism sites have been perceived rather as places of entertainment. From an educational perspective, it would be important to find the most appropriate way to stimulate students' motivation to do more independent research about lighter dark tourism sites as well.

We cannot neglect the role of the product experience instead of 'just a visit to the site' in the context of post-visit effects. The sites with more developed tourism product components, such as offering of guiding services, interactive expositions, and engagement of attendees, obviously have provided more of the learning space. The same factor has influenced higher-level intention for self-reflection or others-directed post-visit effects.

Our study shows visits to dark tourism sites do not often lead to students increasing' professional knowledge about dark tourism. Neither do visits always result in students' desire to explore the topic further. However, field trips often take students to places they have not visited before or would not choose to visit by themselves. This can be a stimulating factor for a general understanding of dark tourism and serve as an eye-opener, as well as provide a number of other benefits – both emotional and cognitive.

Disclosure statement

No potential conflict of interest was reported by the author(s).

Data availability statement

The data that support the findings of this study are available from the corresponding author, upon reasonable request.

ORCID

Ilze Grinfelde ⓘ http://orcid.org/0000-0003-2682-4844

References

Best, M. (2007). Norfolk Island: Thanatourism, history and visitor emotions. *Shima: The International Journal of Research Into Island Cultures, 1*(2), 30–48.

Biran, A., Poria, Y., & Oren, G. (2011). Sought experiences at (dark) heritage sites. *Annals of Tourism Research, 38*(3), 820–841. https://doi.org/10.1016/j.annals.2010.12.001

Chang, H. J., Huang, K. C., & Wu, C. H. (2006). Determination of sample size in using central limit theorem for Weibull distribution. *International Journal on Information and Management Sciences, 17*(3), 31–46.

Chang, L. H. (2017). Tourists' perception of dark tourism and its impact on their emotional experience and geopolitical knowledge: A comparative study of local and non-local tourist. *Journal of Tourism Research & Hospitality, 6* (3). https://doi.org/10.4172/2324-8807.1000169 Retrieved from https://www.scitechnol.com/peer-review/tourists-perception-of-dark-tourism-and-its-impact-on-their-emotional-experience-and-geopolitical-knowledge-a-comparative-study-of-Nz8K.php?article_id=6483

Cohen, E. (2011). Educational dark tourism at an in populo site: The Holocaust Museum in Jerusalem. *Annals of Tourism Research, 38*(1), 193–209. https://doi.org/10.1016/j.annals.2010.08.003

Cooke, S. (2012). Sebald's ghost: Traveling among the dead in the ring of Saturn. In J. Skinner (Ed.), *Writing the dark side of travel* (pp. 47–62). Berghahn Books.

Cooper, M. (2006). The Pacific War battlefields: Tourist attractions or war memorials? *International Journal of Tourism Research, 8*(3), 213–222. https://doi.org/10.1002/jtr.566

Djonko-Moore, C. M., & Joseph, N. M. (2016). Out of the classroom and into the city: The use of field trips as an experiential learning tool in teacher education. *SAGE Open, 6*(2). https://doi.org/10.1177/2158244016649648

Fisher, E. E., Sharp, R. L., & Bradley, M. J. (2017). Perceived benefits of service learning: A comparison of collegiate recreation concentrations. *Journal of Experimental Education, 40*(2), 187–201. https://doi.org/10.1177/1053825917700922

Fokkinga, S. A., & Desmet, P. M. A. (2013). Ten ways to design for disgust, sadness, and other enjoyments: A design approach to enrich product experiences with negative emotions. *International Journal of Design, 7*(1), 19–36. https://doi.org/10.18848/2325-1328/CGP/v07i01/38521

Grinfelde, I., & Veliverronena, L. (2018). The limitations of creative approach: Conducting an orchestra of emotion in the darkness. *Creativity Studies, 11*(2), 362–376. https://doi.org/10.3846/cs.2018.7183

Grinfelde, I., & Veliverronena, L. (2019). Who is guilty that I fail in classroom: Students perspective on higher education. Society. Integration. Education. In *Proceedings of the International Scientific Conference* (Vol. 1, pp. 594–606).

Isaac, R., & Budryte-Ausiejiene, L. (2015). Interpreting the emotions of visitors: A study of visitor comment books at the Grūtas Park Museum, Lithuania. *Scandinavian Journal of Hospitality and Tourism, 4*(4), 400–424. https://doi.org/10.1080/15022250.2015.1024818

Isfrailova, F., & Khoo-Lattimore, C. (2018). Sad and violent but I enjoy it: Children's engagement with dark tourism as an educational tool. *Journal of Tourism and Hospitality Research, 19*(4), 478–487. https://doi.org/10.1177/1467358418782736

Kang, E. J. (2010). Experience and benefits derived from a dark tourism site visit: The effect of demographics and enduring involvement [Unpublished doctoral dissertation]. School of Tourism, The University of Queensland, Queensland.

Kang, E. J., Scott, N., & Timothy, Y. L. (2012). Benefits of visiting a 'dark tourism' site: The case of the Jeju April 3rd Peace Park, Korea. *Tourism Management, 33*(2), 257–265. https://doi.org/10.1016/j.tourman.2011.03.004

Kim, H. J., & Jeong, M. (2018). Research on hospitality and tourism education: Now and future. *Tourism Management Perspectives, 25*, 119–122. https://doi.org/10.1016/j.tmp.2017.11.025

Kirillova, K., Lehto, X. Y., & Cai, L. (2016). Tourism and existential transformation: An empirical investigation. *Journal of Travel Research, 56*(5), 638–650. https://doi.org/10.1177/0047287516650277

Korstanje, M., & George, B. (2017). Emotionality, reason, and dark tourism: Discussions around the sense of death. In M. E. Korstanje & B. George (Eds.), *Virtual traumascapes and exploring the Roots of dark tourism* (pp. 1–26). IGI-Global.

Lankauskas, G. (2006). Sensuous (re)collections: The sight and taste of socialism at Grūtas Statue Park, Lithuania. *The Senses and Society, 1*(1), 27–52. https://doi.org/10.2752/174589206778055682

Lennon, J. J., & Foley, M. (2000). *Dark tourism.* Continuum.

Lennon, J., & Foley, M. (1999). Interpretation of the unimaginable: The US Holocaust memorial Museum, Washington, D.C., and "dark tourism". *Journal of Travel Research, 38*(1), 46–50. https://doi.org/10.1177/004728759903800110

Martini, A., & Buda, D. M. (2018). Dark tourism and affect: Framing places of death and disaster. *Current Issues in Tourism, 23*(6), 679–692. https://doi.org/10.1080/13683500.2018.1518972

Mezirow, J. (1978). Perspective transformation. *Adult Education, 28*(2), 100–110. https://doi.org/10.1177/074171367802800202

Morrison, K. A. (2019). Teaching at dark sites. In K. A. Morrison (Ed.), *Study abroad pedagogy, dark tourism, and historical reenactment: In the footsteps of jack the Ripper and his victims* (pp. 89–115). Palgrave Pivot.

Mortari, L. (2015). Emotion and education: Reflecting on the emotional experience emotion and education. *European Educational Research Journal, 4*(4), 157–176. https://doi.org/10.12973/eu-jer.4.4.157

Pattachini, L. (2018). Experiential learning: The field study trip, a student-centred curriculum. *Compass: Journal of Learning and Teaching, 11*, 2. https://doi.org/10.21100/compass.v11i2.815

Sather-Wagstaff, J. (2011). *Heritage that hurts: Tourists in the memoryscapes of September 11.* Left Coast Press.

Seaton, A. (1996). Guided by the dark: From thanatopsis to thanatourism. *International Journal of Heritage Studies, 2*(4), 234–244. https://doi.org/10.1080/13527259608722178

Shackley, M. (2001). Potential Futures for Robben Island: shrine,museum or theme park? *International Journal of Heritage Studies, 7*(4), 355–363.

Sharpley, R., & Baldwin, F. (2009). Battlefield tourism: Bringing organized violence back to life. In R. Sharpley & P. R. Stone (Eds.), *The darker side of travel: The theory and practice of dark tourism aspect of tourism Series* (pp. 186–206). Channel View Publications.

Stone, P. R. (2006). A dark tourism spectrum: Towards a typology of death and macabre related tourist sites, attractions and exhibitions. *Tourism, 54*(2), 145–160.

Stone, P. R. (2012). Dark tourism and significant other death: Towards a model of mortality mediation. *Annals of Tourism Research, 39*(3), 1565–1587. https://doi.org/10.1016/j.annals.2012.04.007

Tal, T., & Morag, O. (2009). Reflective practice as a means for preparing to teach outdoors in an ecological garden. *Journal of Science Teacher Education, 20*(3), 245–262. https://doi.org/10.1007/s10972-009-9131-1

Timm Knudsen, B. (2011). Thanatourism: Witnessing difficult pasts. *Tourism Studies, 11*(1), 55–72. https://doi.org/10.1177/1468797611412064

Tyng, C. M., Amin, H. U., Saad, M. N. M., & Malik, A. S. (2017). The influences of emotion on learning and memory. *Frontiers in Psychology, 8*, 1454. https://doi.org/10.3389/fpsyg.2017.01454

Walker, J., & Manjamba, V. N. (2020). Towards an emotion-focused, discomfort-embracing transformative tourism education. *Journal of Hospitality, Leisure, Sport & Tourism Education, 26*. https://doi.org/10.1016/j.jhlste.2019.100213

Walter, T. (2009). Dark tourism: Mediating between the dead and the living. In R. Sharpley & P. R. Stone (Eds.), *The darker side of travel: The theory and practice of dark tourism aspect of tourism series* (pp. 39–88). Channel View Publications.

Wang, H. H., & Carlson, S. P. (2011). Factors that influence student's satisfaction in an environmental field day experience. *International Electronic Journal of Environmental Education, 1*(2), 129–139.

Yan, B. J., Zhang, J., Zhang, H. L., Lu, S. J., & Guo, Y. R. (2016). Investigating the motivation-experience relationship in a dark tourism space: A case study of the Beichuan earthquake relics, China. *Tourism Management, 51*(1), 108–121. https://doi.org/10.1016/j.tourman.2015.09.014

Yoshida, K., Huong, T. B., & Lee, J. T. (2016). Does tourism illuminate the darkness of Hiroshima and Nagasaki? *Journal of Destination Marketing & Management, 5*(4), 333–340. https://doi.org/10.1016/j.jdmm.2016.06.003

Dark tourism as educational tourism: the case of 'hope tourism' in Fukushima, Japan

Kyungjae Jang ⓘ, Kengo Sakamoto and Carolin Funck

ABSTRACT

This article examines how dark tourism has been adopted and developed in Japan and suggests how it can be linked to educational tourism, through the case of a 'Hope Tourism Guided Tour' held in Fukushima in 2018. In Japan, dark tourism has developed into a new form, nested within educational tourism (or an educational tourism that contains elements of dark tourism within it). This is connected to the long history of educational tourism and to the cultural backlash against dark tourism. Hope tourism in Fukushima is an example of this type of new educational tourism-centred dark tourism. First, we review how the concepts of educational and dark tourism have spread in Japan. Second, the article clarifies how educational dark tourism has developed in the case of the Fukushima Hope Tourism Guided Tour.

Introduction

Dark tourism, a concept proposed by Lennon and Foley (1996, 2000), defined as travelling to places historically associated with death and tragedy, has stimulated a variety of discussions since its introduction to Japan around 2007. The discussions have addressed common issues related to dark tourism as pointed out by Stone (2011). Stone discusses the moral issues, media, propaganda issues, interpretations, political issues, management, operational issues, and sociocultural and sociological issues related to the controversies surrounding dark tourism.

Since it is conducted in and around places associated with a memory of death, dark tourism often emphasizes its educational purpose (Foley and Lennon, 1996, 2000; Israfilova & Khoo-Lattimore, 2019; Stone, 2011; 2012). Stone (2012) emphasized the role of dark tourism as an educational element in tourism to New York's Ground Zero and emphasized that it can increase engagement.

Meanwhile, dark tourism in Japan and its use for educational purposes are somewhat different from those in the West. Contrary to the Western idea that dark tourism per se is used as an educational tool, in Japan, rather than branding destinations as 'dark tourism,' there are many cases in which an element of dark tourism is included in educational tourism.

Discussions of dark tourism in Japanese society are related to basic perceptions of tourism and 'darkness.' Although the interpretation is somewhat metaphorical, the Japanese word *kankō* (tourism), means 'seeing the light.' It is said to be derived from 'seeing the light of the country' in the Chinese classic the *I Ching* (Book of Changes). In other words, the term 'dark tourism' can be interpreted to overturn the Japanese concept of tourism as 'seeing the light.' On the other hand, Japan also has a cultural phenomenon called *fukinshin* (which roughly translates as 'indiscreet' or 'inappropriate'). When a disaster or calamity occurs, sympathy from the whole of society is required, and pleasant or enjoyable activities at these times are considered to be inappropriate (Makihara, 2011; Okuno, 2013). The word is often used in connection with social empathy, such as when the Great East Japan Earthquake occurred on 11 March 2011.

Tourism in Japan, meanwhile, also emphasizes aspects of education. School trips, which are official tours conducted by secondary schools for educational purposes, are the most representative example of this. The school trip system, which began in the Meiji period (1868–1912), remains the paradigm of educational tourism in modern Japan and is also seen in Taiwan and Korea, which were influenced by the Japanese education system in the colonial age. Educational tourism is also related to the Japanese social system which emphasizes the importance of work. Thus, in Japanese society, the development of tourism is a way to pursue work and play together (Horino, 2017).

Based on this awareness of dark tourism and the tradition of educational tourism, Japan has developed a combination of the two, of which representative examples are Hiroshima and Fukushima. Hiroshima, which was bombed in 1945, is a typical destination for dark tourism (Stone et al., 2018). In Japan, however, Hiroshima is not considered to be a destination for dark tourism, as the use of the term 'dark tourism' has the potential to be *fukinshin* for the victims. Hiroshima should be considered rather a symbol of peace and a place of commemoration. For that reason, peace tourism, a form of dark tourism with an added element of education, is being developed in Hiroshima City (Hiroshima Peace Tourism, n.d..). The situation is similar in Fukushima, which has been called the second Chernobyl due to the Fukushima Daiichi Nuclear Power Plant disaster that occurred during the 2011 Great East Japan Earthquake and following a tsunami. Fukushima Prefecture also avoids the term dark tourism. For example, a journalist from New Zealand visited the Fukushima region to film the second episode of the Netflix series 'Dark Tourist.' Immediately after the broadcast, Fukushima Prefecture came forward and announced it was considering legal action against the show's producers (NZ Herald, 2018). Similar to Hiroshima, instead of dark tourism, Fukushima Prefecture focuses on a form of educational tourism called 'hope tourism.'

Peace tourism and hope tourism are alternatives that avoid the term 'dark tourism' due to the cultural background of *fukushin* but still include elements of dark tourism within a context of educational tourism, a situation which can provide new insights into the discussion of dark tourism. Rather than simply expanding the study of dark tourism geographically to the Japanese context, this study makes it possible to find out how dark tourism combines with educational tourism in Japan and what new experiences can be provided to tourists in the process.

This article analyses the formation of new kinds of tourism in Japan, specifically the combination of dark tourism with educational tourism, through the case of hope tourism in Fukushima. First, the article reviews how the concepts of educational tourism and dark tourism were created, developed, and disseminated in Japan. Second, the article clarifies how educational dark tourism in Japan is developed through the case of Fukushima hope tourism. Specifically, the authors analyse how dark tourism functions and what educational effects it has, based on a survey of the participants of a 'Hope Tourism Guided Tour' held in Fukushima Prefecture on 27–29 October 2018 (Saturday to Monday).

Educational tourism and dark tourism in Japan

Educational tourism in Japan

Attention to the educational aspects of tourism is not new, whether in popular media or academically. Previous studies have proposed the concepts of educational travel (Kalinowski & Weiler, 1992) and edu-tourism (Holdnak & Holland, 1996). Smith and Jenner (1997) described the concept of educational tourism as a leisure-education hybrid. As a systematic definition of tourism, Ritchie et al. (2003) define educational tourism as 'a tourist activity undertaken by those for whom education and learning is a primary or secondary part of their trip' (p. 18). In other words, all forms of tourism with elements of learning can be called educational tourism. Funck (2008) summarizes learning in tourism (Figure 1), based on the characteristics of the place, the motive of tourism and the tourist's degree of freedom.

School trips are a representative form of Japanese-style educational trips mainly conducted in secondary education, where students in all grades go to the same destination for several days. Recently, however, the scale has been reduced: trips tend to be organized by individual classes rather than all grades together. Japan's school trip system began in 1888 and developed during the Taisho period (1912–1926) as part of nationalist education, in which students visited military facilities or Japanese colonies such as the Korean peninsula (Hoshino, 1997). Post-war school trips since then mainly visit historical sites and destinations associated with peace, but the educational purpose remains.

Based on this tradition, the Japan National Tourism Organization also began promoting inbound educational tourism, where foreigners visit Japan for educational purposes. The Japan National Tourism Organization (n.d.) introduces inbound educational tourism to Japan as follows:

> Educational travel refers to group tours organized by schools for their students with faculty members as group leaders. These trips have clear learning objectives and usually include visits to local schools, site visits, hands-on activities, and so much more.

This can be said to be a kind of special interest tourism aimed at learning. Based on the long tradition of school trips, it serves as a strategy to attract inbound tourism through educational tourism combined with other forms of tourism such as dark tourism. The Fukushima hope tourism covered in this article is one example of tourism that combines elements of dark tourism and educational tourism.

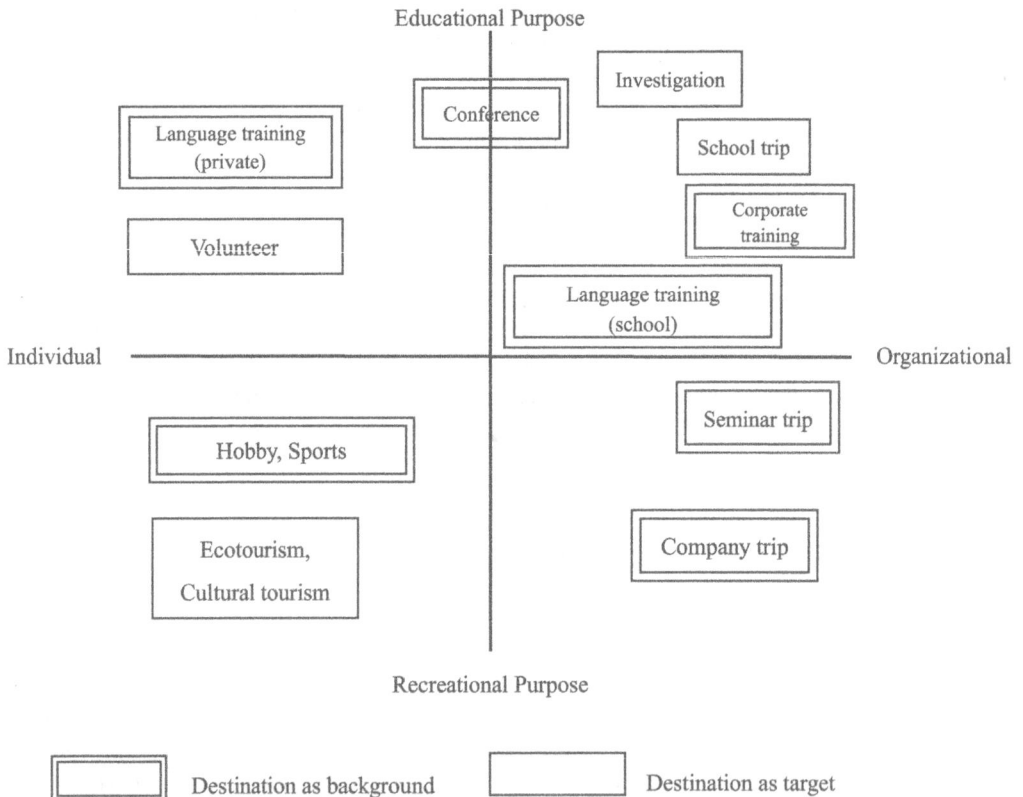

Figure 1. Educational tourism (Funck, 2008, p. 162).

Dark tourism in Japan

Dark tourism in Japan was first introduced by tourism geographer Carolin Funck, circa 2007 (Funck, 2008). Dark tourism, which had not been seen in the Japanese tourism industry and society up to that point, emerged after the 2011 Great East Japan Earthquake on 11 March 2011. Among other accidents, the Fukushima Daiichi Nuclear Power Plant disaster, which carried with it the image of the Great East Japan Earthquake and was classified as one of only two Level 7 events in world nuclear history (along with Chernobyl in 1986), played a large role.

The most important task in Japanese society after the earthquake and tsunami, including the nuclear accident, was to find ways to rebuild and revive the affected region. Revival plans were offered from across society, including academia, and dark tourism was also highlighted. Writer and philosopher Hiroki Azuma formed the Fukuichi Kankō Project (Fukushima Daiichi nuclear plant tourism project) in 2012 with journalist Daisuke Tsuda and sociologist Hiroshi Kainuma. The purpose of this project is to turn the disaster site of the Fukushima Daiichi Nuclear Power Plant into a tourist destination. They visited Chernobyl in 2013 and published a book, *Tourizing Fukushima: The Fukuichi Kanko Project* (Azuma, 2013). In this book, tourism researcher Akira Ide introduced the concept of dark tourism, which led to an active discussion of dark tourism in Japan.

Subsequently, dark tourism in Japan has been discussed in various ways, including its aspects of learning, memorialization, and voyeurism. Ide (2013, p. 145) considers mourning and inheriting regional sorrow to be the core of dark tourism. Although he did not emphasize it directly, this argument supports the importance of the educational element in dark tourism. Ichinosawa (2016) focuses on tourists' inherent 'pleasure at glimpsing the suffering of others' (p. 52) and states that 'pleasure in darkness' is at the core of dark tourism. For example, suicide tourism (visiting a country where euthanasia is legal), is not discussed in the context of dark tourism despite being tourism related to death and sadness (pp. 51–52). Furuichi (2015), who visited war museums around the world, also asks rhetorically 'How much I've come to face the horrors of the war so seriously, dark tourism is fun and exciting' (p. 92). From the perspective of the desire to consume tragedy, dark tourism contains the idea of the darkness of pleasure.

Some of the debates about dark tourism emphasize the value of seeing not only 'light' but also 'shadows' (negative aspects) to enable a deeper understanding of the destination. Fukami (2017) defines dark tourism as tourism that aims to look directly at 'shadows' as well as 'light' (p. 186). On the other hand, Ichinosawa (2016) regards 'touching' the suffering of others as the principal value of dark tourism. He emphasizes that the meaning of tourism is to satisfy the fundamental needs of humans, rather than seeing tourism as a way to approach society and social issues as claimed by Fukami (2017) and Ide (2018). Sudo (2016) considers dark tourism to be a consumption behaviour for modern people who have lost an understanding of life and seek the reality of life through a confrontation with death. In other words, in dark tourism, sightseeing behaviour is aimed at entertainment rather than education. From this perspective, dark tourism is more likely to be considered and combined with entertainment.

However, in reality, especially in Japan, local governments in particular tend not to use the term dark tourism. They worry that although dark tourism contains both 'light' and 'shadow,' if the destination is described as dark tourism, it will literally take on a negative image. Ide (2013, p. 148) pointed out that the concept of dark tourism has not penetrated Japan due to the poor wording using 'dark'; Kainuma (2013) and Ide (2018) suggest that in the case of Fukushima, the terms hopeful tourism or reconstruction tourism would be better for the region. In addition, in 2018, content about Fukushima considered *fukinshin* (inappropriate) was distributed globally in the Netflix documentary *Dark Tourist*. This created an image of dark tourism as a dishonest form of tourism that consumes the tragedies of others.

It is natural that locals do not like it when the image of their area is said to be 'dark.' However, wherever humans have lived, there must have been death or sadness. In that sense, tourism aiming at incorporating shadow, not just light, is an effective method for attracting tourist interest. However, as the concept of 'pleasure in darkness' suggests, it is always necessary to keep in mind that a guest's gaze at the shadow can be felt as a form of violence by the host community.

Fukushima hope tourism

Overview

The impact of the earthquake and tsunami on Fukushima Prefecture's tourism has been significant. The number of tourists in Fukushima Prefecture was 57 million in 2010 but dropped by 39.4% in 2011 due to the disaster. Since then, it has been recovering, except for 2014, which showed a slight decrease from the previous year. According to the latest 2017 data, it has reached 95.4% of the 2010 level (Fukushima Prefecture Commerce and Industry Labor Relations Tourism Exchange Bureau Tourism Exchange Division, 2018). However, according to nationwide data, the number of tourists in 2017 was 120.6% of the level in 2010. By region in Fukushima, Nakadori and Aizu surpassed pre-earthquake levels, while Hamadori, where 'difficult-to-return' areas remain, stayed at 68.1% of 2010 levels.

In this situation, Fukushima Prefecture started the hope tourism project in 2017, which is intended to be the antithesis of dark tourism. The prefecture defines hope tourism as:

> In Fukushima Prefecture, we are promoting 'hope tourism', a journey that will make you grow as you witness Fukushima as it stands now and encounter the people who continue to work towards its reconstruction, making you think about the lessons from the earthquake and nuclear accident, reconstruction, and overcoming adversity. Fukushima Hope Tourism (n.d.)

Hope tourism is intended for junior and senior high school students in Japan, foreign students studying in Japan, and overseas news organizations. The prefecture has already conducted tours for junior and senior high schools in Japan and will promote it as an educational tourism programme titled Fukushima Gasshuku (Fukushima field study).

In this context, the prefecture prepared a guided tour for international students to widen the concept of hope tourism from national to international. Hiroshima University's 13 international students and two Japanese students participated in the guided tour held from 27 to 29 October (Saturday to Monday), 2018. Participants were students from the School of Integrated Arts and Sciences, Department of Integrated Global Studies, and the Graduate School of Taoyaka Programme (Table 1). Participants visited a total of nine places, including Tokyo Electric Power Decommission Archive Centre and a local community centre, where they could learn about earthquakes, tsunamis, and nuclear power plants. The programme also included trips to enjoy local festivals such as a wild sesame harvest festival.

Table 1. Participants in the Hope Tourism 2018 case study.

No.	Country of origin	Gender	Age	Length of stay in Japan (years)
A	Taiwan	M	20	4
B	Thai	M	18	0.5
C	China	F	34	5
D	Taiwan	F	28	1
E	China	M	26	3
F	China	F	25	3
G	China	F	21	2
H	Vietnam	M	39	4
I	Vietnam	F	33	11
J	Vietnam	F	35	1.5
K	Indonesia	M	25	1
L	Myanmar	F	25	1
M	Indonesia	F	29	4
N	Japan	F	22	22
Author (Sakamoto)	Japan	M	22	22

The tour was hosted by the Fukushima Prefecture Tourism & Local Products Association, but the Fukushima Prefectural Government Tourism Promotion Division and travel agencies were also involved. In hope tourism, a 'field partner' accompanies the entire journey. Participants listen to explanations from guest speakers at each destination while obtaining information from the field partner. Tour courses and routes are tailor-made for each participating organization. The theme of 2018 for Hiroshima University students was Hiroshima (peace) and Fukushima (hope), and the field partner this time was William McMichael, assistant professor at Fukushima University (Figure 2). The course of the tour is the areas shown in Figure 2, consisting of a waste landfill information centre located north and south of the Fukushima Daiichi Nuclear Power Plant, a harvest festival at a nearby farm, field study, and exchange with residents. It is a tour that combines learning and experiences, where participants can get information on nuclear power plant waste and interact with local residents who are otherwise difficult to access.

Method

A total of six surveys were conducted before, during and after the tour on the tour participants (Table 2). The survey was conducted in a mixture of Japanese and English, and the tour itself was conducted in English. However, it should be noted that none of the participants was a native

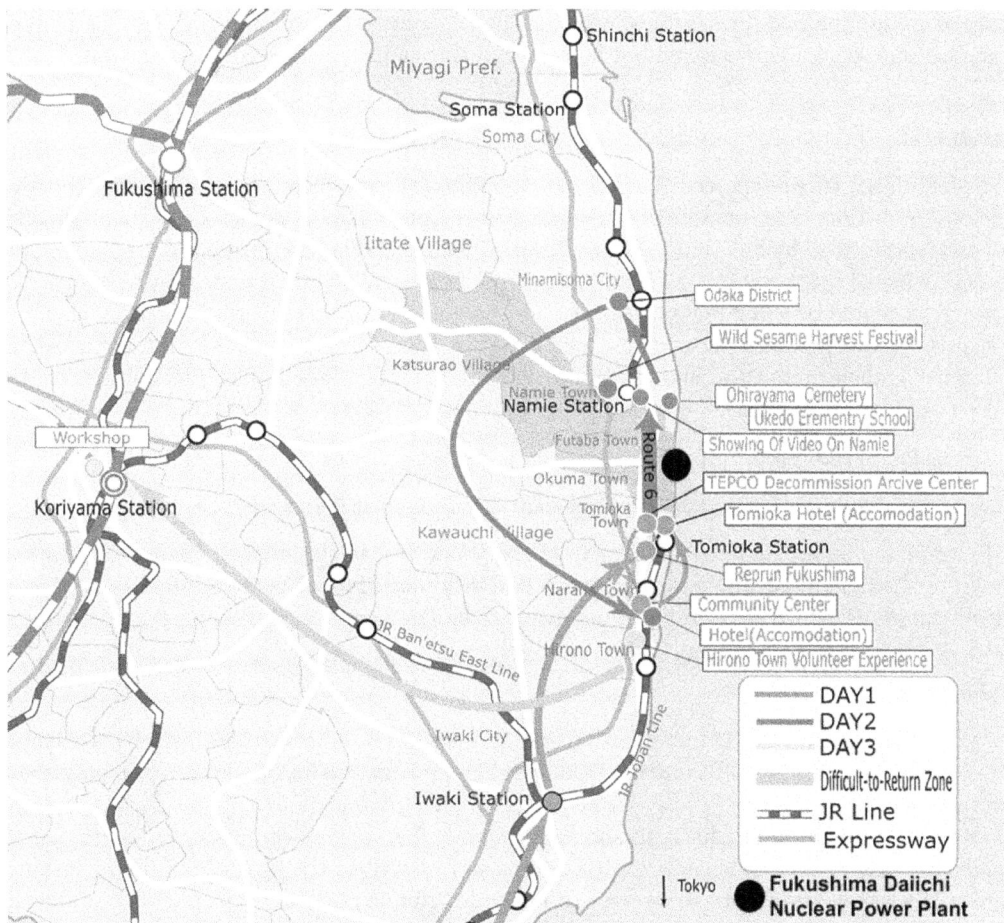

Figure 2. Tour area for Hope Tourism 2018 (Courtesy: Fukushima Prefecture).

Table 2. Surveys of Hope Tourism 2018 case study.

Preliminary survey	(i)	Pre-questionnaire	The image and motivation for participating, and expectations of the tour
	(ii)	Group interview	Based on pre-questionnaire
Guided tour	(iii)	Field note by author	Information provided during the tour, observation of participants
Post survey	(iv)	Post-questionnaire	About satisfaction and willingness to revisit
	(v)	Travel diary	Impressions of tour
	(vi)	Group discussion	About the future of Fukushima tourism

speaker of English and the degree of their command of English varied greatly. The pre-survey asked the motivation for participation, the image of Fukushima, and the awareness of disasters, and the post-survey asked about the degree of satisfaction of each trip, changes in the image of Fukushima, and willingness to revisit, to try to understand the participants' motivations.

Specifically, the following categories were used to measure the element of education or 'shadow': information and knowledge of disaster and disaster prevention; information and knowledge of post-disaster reconstruction; mourning and prayer for the victims of disaster; and scenery of the disaster-stricken area.

On the other hand, the following items were used to measure the elements of pleasure or 'light': natural scenery, food and specialties, and culture and sightseeing attractions.

As a specific method of data analysis, i and iv collected data using Google Forms, an internet survey tool. For iv, questionnaires were distributed and answered in an additional way. A five-stage questionnaire was created to determine respondents' motivations for participation, the image of Fukushima, and the degree of awareness of the damage. Furthermore, through the analysis of the free questionnaire, we found out how participants actually perceived hope tourism.

Results

The preliminary survey showed that 13 of 14 responded that their image of Fukushima was 'nuclear,' 'radiation,' and 'nuclear power plant.' On the other hand, 'tsunami' was mentioned by two and 'earthquake' by only one. The initial recruitment for this tour comprised a group of students who hoped to learn about the disaster, so it may be natural that their answers were connected. However, it should be noted that nuclear disasters have a more salient image than earthquakes and tsunamis. At least for these participants, their image of Fukushima was that of a nuclear accident.

Regarding their motivation, most respondents said that they wanted to learn about the disaster and reconstruction, such as:

I would love to know about Fukushima and how the people living after the nuclear accident. (H)

Want to experience the revitalization of the Fukushima and mindset of the local people about the accident. (L)

Based on the survey answers, participants talked about their image of and interest in Fukushima at a group discussion. The main topic here was that they wanted to see with their own eyes the reality of Fukushima, which has only become possible recently. There were interesting remarks such as:

There was a lot of news about mysterious creatures coming out of the nuclear disaster. I want to see if this is true. (F)

Meanwhile, analysis of the follow-up surveys and the travel diaries written by the participants revealed how they actually felt about the Fukushima Hope Tour and how their perceptions changed before and after.

The first, as an aspect of education and dark tourism, the Tokyo Electric Power Decommission Archive Centre, which participants visited on first day, described the situation of the Fukushima Daiichi Nuclear Power Plant. There were so many descriptions that the explanation was difficult to understand. It seemed technically too advanced for the participants. Furthermore, there were many descriptions of how much the situation had improved immediately after the accident, but little discussion about the future. Topics such as 'how to extract fuel debris in the future' and 'how to treat treated water containing tritium in the future' were only included in the final question and answer session, and there were no specific comments about the roadmap for decommissioning. Participant (C) wrote in her travel diary: 'The important part was skipped and I wanted to know how to dispose of the water in the tank and the fuel left in that building.'

Another place related to the earthquake is Ukedo Elementary School (Figure 3), which participants visited on day 2. The elementary school building has not been demolished since the tsunami and remains standing now. It was often described as a successful example of evacuation for the children and teachers who were at school at the time. In addition, this school building was previously left open, so that visitors could leave messages of support and encouragement on blackboards and notebooks until some inappropriate graffiti was found. In response, participants commented on the horror of the tsunami and the state of what remains:

> I was deeply moved by [the fact that] teachers in Ukedo Elementary School were escaped safely on the biggest tsunami in 2011. (J)

> I believed that Ukedo Elementary School will be one of the most wonderfully miracle tourist spot ever in the world in the near future. (B)

At the same time, participants also had the opportunity for pleasurable experiences. Before Ukedo elementary school, participants visited a wild sesame harvest festival held close to the difficult-to-return zone in Namie Town, which celebrated the harvest of wild sesame after the evacuation order was rescinded. Although it was a small local event, participants left comments that suggested it could be considered a symbol of revival.

Figure 3. Ukedo elementary school and participants of Hope Tourism 2018. Photo taken by Kengo Sakamoto.

I think it's the warmest festival I've ever attended in Japan. (F)

I love the local products. Those are not only economic benefits but also alive supports for the hearts of the residents. (I)

On the last day, they had the opportunity to participate directly in revival activities. As part of the regional reconstruction, cherry trees are planted along Route 6 (a main road to Fukushima), and participants volunteered to mow the surrounding area. They did just under an hour of work, but many commented that it was very enjoyable, and were satisfied that they had the opportunity to actually work for the region.

I was so glad I felt a little helpful. (F)

By doing that, it would be more memorable for the participants. (M)

Subsequently, we codified the travel diaries the participants had written during the tour (Table 3). Naturally, due to the nature of study tours, there were many descriptions of the disaster and reconstruction. Individual codes with the largest number were 'warm' and 'delicious,' indicating that the local atmosphere and meals experienced at the sesame harvest festival and similar opportunities left a strong impact.

Finally, results for satisfaction with the tour and change in perception of Fukushima were positive. In the post-hoc questionnaire, overall satisfaction with the guided tour was 4.79 out of 5. The reasons for this high satisfaction do not only relate directly to the recollections of the disaster they experienced, but also other appealing aspects and possibilities for learning.

I had many experiences about the extreme disaster in 2011, the local people, the foods and the beautiful landscapes there. (J)

The places we visited are really inspiring for me, full of stories, rich of hope and lots of knowledge about social, technical and cultural aspects. (L)

After joined the tour, I had different perspective toward Fukushima in a more positive ways. (M)

The change in image of Fukushima as a result of the tour was 4.79 out of 5 points (1 = 'no change in image of Fukushima' to 5 = 'complete change in image'). The reasons for the evaluation are:

Different than what is written on the Internet. (A)

Fukushima is not a scary place; it has a good economy and there are delicious things, so it is a good tourist destination. (C)

Table 3. Codes from travel diaries of Hope Tourism 2018 participants.

Aspect	Topics	Codes
Studying (65)	Information and knowledge about disaster and disaster prevention (21)	Sadness (5), advanced and difficult (5), easy to understand (5), speechless (4), impressed (2)
	Information and knowledge on post-disaster reconstruction (33)	Efforts of locals (6), strong (5), change of image (4), significance of participation in activities (3), reduction of air radiation dose (3), possibility of tourism industry (2), impression (2), bright atmosphere (2), positive thinking (2), intellectual curiosity (2), consent to participate (1), selfishness (1)
	Memorial/prayer for victims of disaster (0)	
	Scenery of the stricken area seen with their own eyes (11)	Shock (6), lonely (3), scale of tsunami (1), scary (1)
Pleasure (14)	Landscape rich in nature (3)	Landscape of Cape Tenjin (3)
	Unique food and speciality products of Fukushima (8)	Delicious (7), intellectual curiosity (1)
	Unique culture and attractions of Fukushima (3)	Hot spring (3)
Communication (21)	Story of local people, dialogue with local people (21)	Warm atmosphere (8), universal message (4), rest (4), hospitality (3), hesitation (2)

Discussion and conclusions

This article has examined how dark tourism has been adopted and developed in Japan and suggests how it can be linked to educational tourism through the case of Fukushima hope tourism. The major characteristic of Fukushima hope tourism is that it does not bring dark tourism to the fore, but is a type of tourism that focuses on educational tourism that contains elements of dark tourism. There, the social and cultural context of Japan and the context of the politicization of places are working together.

Fukushima hope tourism is strictly operated by the local government. This originates from Japan's educational tourism tradition, which has a long history of organizing school trips. Hope tourism has a structure in which participants do not only see what they want to see but see what they are shown. In other words, rather than actively satisfying desires through tourism, it is a system that is likely to create passive recipients. To compensate for this, the schedule is tailor-made according to the participating organizations, but the participants' passiveness during the actual tour has not been addressed. As for the role of the guide, although they are essential in improving educational effectiveness, the intervention of full-time guides may limit the freedom of participants and their information acquisition and communication. What is important here is the presence and role of the intermediary who leads the hope tourism and connects the region and the participants. For tourism that includes educational elements, the effects of individual trips are limited. Fukushima hope tourism selects an appropriate guide for each tour and they accompany the entire itinerary. The guide is both an interpreter and a negotiator. The guide conveys information to the participants according to the overall theme of the tour and the meaning of the place. In addition, the local residents were informed about the purpose of participants and the tour and were provided appropriate guidance. Participants hear locals tell their stories directly during the tourism experience, which is enhanced by the guide's prior coordination.

On the other hand, it also can be said that Fukushima is being used politically to serve a logic that gives priority to educational tourism. It is necessary for the region and the Japanese government to emphasize that the risk of the region surrounding the nuclear power plant is sufficiently low, which inevitably leads to appropriate control of participants and training in the field. Fukushima became a symbol of danger immediately after the earthquake. The dangers of nuclear power, radioactive spills, ruined villages, and endless piles of contaminated soil have made Fukushima a destination for dark tourism. However, the Japanese government urgently needed to overcome the disaster and create a safe area, especially in order to create a secure and reassuring symbol of Japan ahead of the 2020 Olympics. In this situation, Fukushima has been used as a symbol of revival and safety. In the process, selective work was carried out to prevent forgetting of place and to pass on memories. In the Fukushima hope tourism, the elements of death and danger are limited, and educational elements are introduced concerning safety efforts and memories of survival. In the reactions of the participants, the nuclear power plant, the problem of polluted water treatment, the current potential risks, and future problems were seen to require a high degree of scientific knowledge and were not fully understood by experts, due to the unpredictable future surrounding the nuclear plant. Nevertheless, sharing and exchanging local challenges can visualize invisible risks and provide opportunities to work together to find solutions. Furthermore, these attempts have the advantage of ensuring regular income in areas where revival is needed.

Another important issue here is how to combine pleasure in this type of tourism with dark tourism and education. As mentioned above, understanding the educational elements of Fukushima requires a high degree of natural science and social science literacies. This is also a factor that undermines the public's willingness to visit Fukushima. It creates an image that the place is not easily accessible to the general public and is difficult to visit. Being able to provide an element of pleasure acts as a motivation here. As seen in the reflections of the hope tourism participants, experiences of local festivals and volunteering could offset the confusion caused by the educational aspects. This element of pleasure can be an incentive to draw participants into the educational element. By combining the two elements, participants can understand lessons from disasters more effectively and increase their interest in the region.

In areas that have experienced large-scale disasters such as Fukushima, the social and economic structure of the entire area is reorganized. This also applies to the tourism sector. In the course of discussions about using tourism as a device for the preservation and shaping of memory and economic development, the necessity of various forms of tourism has been discussed, including dark tourism, educational tourism, and tourism for pleasure. Fukushima hope tourism has emerged as the first dark tourism antithesis, though it is still in development. The analysis of the guided tour gave insight into how education and enjoyment can be combined, and how the two can complement each other to help better understand and actually promote a region. If the programme is merely a propaganda tool for revival, it will have a negative impact on the participant's motivation and willingness to revisit, and merely pursuing pleasure, participants will not gain an understanding of the area and may provoke local opposition.

Acknowledgements

This article is dedicated to all those working to commemorate the memory of the 2011 Great East Japan Earthquake and tsunami.

Disclosure statement

No potential conflict of interest was reported by the author(s).

Funding

The Hope Tourism Guided Tour is sponsored by Fukushima Prefecture.

ORCID

Kyungjae Jang http://orcid.org/0000-0001-8650-2324

References

Azuma, H. (2013). *Tourizing Fukushima:The Fukuichi Kanko project. Shiso-Chizu (idea map) β vol.4-2*. Genron.

Foley, M., & Lennon, J. J. (1996). JFK and dark tourism: A fascination with assassination. *International Journal of Heritage Studies*, 2(4), 198–211. https://doi.org/10.1080/13527259608722175

Fukami, S. (2017). Hidden Christian-related heritage in Nagasaki and Amakusa regions and dark tourism: From the viewpoint of the encounter between guest and host. *Tourism Studies Review*, 5(2), 185–196.

Fukushima Commerce and Industry and Labor Department Tourism Exchange Bureau Tourism Exchange Division. (2018). *Fukushima Prefecture Tourist Status Year 2017*.

Fukushima Hope Tourism. (n.d.). *Concept.* https://www.hopetourism.jp/en/concept.html

Funck, C. (2008). 'Learning tourism' and the creation of regional knowledge. *Chiri-kagaku (Geographical Sciences), 63*(3), 160–173. https://doi.org/10.20630/chirikagaku.63.3_160

Furuichi, N. (2015). *Nobody can teach war.* Kodansha+α Bunko.

Hiroshima Peace Tourism. (n.d.). *Hiroshima peace tourism.* https://peace-tourism.com/en/top.html

Holdnak, A., & Holland, S. (1996). Edu-tourism: Vacationing to learn. *Parks and Recreation, 31*(9), 72–75.

Horino, M. (2017). 'Seriousness' and 'play' in tourism II. *Nara Prefectural University Kenkyu Kiho Quarterly Review, 27*(4), 87–94.

Hoshino, A. (1997). The history of school excursions: Prewar part. *Geographical Education, 26*, 6–15.

Ichinosawa, J. (2016). Darkness of amusement: Dark tourist experience and attractiveness of dark tourism in the cases of disaster memorial parks and museums. *Memoirs of Institute of Humanities, Human and Social Sciences, Ritsumeikan University, 110*, 23–60.

Ide, A. (2013). Considering dark tourism. In H. Azuma (Ed.), *Tourizing Fukushima: The Fukuichi Kanko project. Shiso-Chizu (idea map) β vol.4-2* (pp. 144–157). Genron.

Ide, A. (2018). *Expansion of dark tourism.* Bijutsusppansha.

Israfilova, F., & Khoo-Lattimore, C. (2019). Sad and violent but I enjoy it: Children's engagement with dark tourism as an educational tool. *Tourism and Hospitality Research, 19*(4), 478–487. https://doi.org/10.1177/1467358418782736

Japan National Tourism Organization. (n.d.). *Discover & learn Japan.* https://education.jnto.go.jp/en/discover-learn-japan/

Kainuma, H. (2013). Touristification from Now On. In H. Azuma (Ed.), *Tourizing Fukushima: The Fukuichi Kanko project. Shiso-Chizu (idea map) β vol.4-2* (pp. 54–57). Genron.

Kalinowski, K. M., & Weiler, B. (1992). Educational travel. In B. Weiler & C. M. Hall (Eds.), *Special interest tourism* (pp. 15–26). Belhaven.

Lennon, J. J., & Foley, M. (2000). *Dark tourism.* Continuum.

Makihara, K. (2011, March 15). *Japanese stiff upper lip: Please avoid fukinshin.* TIME. http://content.time.com/time/world/article/0,8599,2059164,00.html

NZ Herald. (2018, September 3). *Japanese authorities consider action over David Farrier's dark tourist.* NZ Herald. https://www.nzherald.co.nz/entertainment/news/article.cfm?c_id=1501119&objectid=12118229

Okuno, K. (2013, December). The Japanese OTAKUs' one month activity just after the 3.11 earthquake. In *2013 International Conference on Signal-Image Technology & Internet-Based Systems* (pp. 401–407). IEEE.

Ritchie, B. W., Carr, N., & Cooper, C. P. (2003). *Managing educational tourism.* Channel View Publications.

Smith, C., & Jenner, P. (1997). Educational tourism. *Travel & Tourism Analyst, 2*(3), 60–75.

Stone, P. R. (2011). Dark tourism: Towards a new post-disciplinary research agenda. *International Journal of Tourism Anthropology, 1*(3–4), 318–332. https://doi.org/10.1504/IJTA.2011.043713

Stone, P. R. (2012). Dark tourism and significant other death: Towards a model of mortality mediation. *Annals of Tourism Research, 39*(3), 1565–1587. https://doi.org/10.1016/j.annals.2012.04.007

Stone, P. R., Hartmann, R., Seaton, A. V., Sharpley, R., & White, L. (2018). *The Palgrave Handbook of dark tourism studies* (pp. 335–354). Palgrave Macmillan.

Sudo, H. (2016). Modernity and ambiguity of dark tourism (special issues: Dark tourism). *Memoirs of Institute of Humanities, Human and Social Sciences, Ritsumeikan University, 110*, 85–109.

Conclusion: future research directions

Rami K. Isaac

Despite the widespread of the concept of dark tourism, which has provided, for a generation of researchers and practitioners, a stimulus to many by opening up many new areas of research, it is clear that the relationships between tourism and death/suffering are a valid theme within tourism research (Light, 2017). Nevertheless, several authors (Biran & Poria 2012; Bowman & Pezzullo, 2010) have questioned whether the concept has become so broad and multi-faceted and whether it is meaningful pursing the dark tourism thought, which is problematic as a name. In the following section, there are two future research priorities identified, namely the experience and emotions of visitors to sites associated with death and suffering.

Experience

The experience is something that happens to somebody as an exposure to a circumstance. This may be an experience of the *tourist/visitor, a non-tourist/non-visitor, or bystander*. These could be heritage experience (if the site is part of the visitors' heritage), in this case, I would call it as (*non-tourists/non-visitor*) (who feel bound in some ways to the site may experience emotions as a result of their knowledge, observation or imagination of the tourist experience), educational cultural experience (for those who wish to gain knowledge or for those whose social agents think they should) as well as other tourist experiences (e.g., those visiting only as the site is a must-visit tourist attraction), which is basically experienced by the *tourists*. We can, in addition, differentiate between different types of visitors such as *victims, perpetrators, and bystanders*. *Victims* are those who identify themselves in some ways with the victims, which results in action to obtain justice or reparation for the victims of perceived injustice. *Perpetrators* feel some links with those who committed the atrocity even at a distance. *Bystanders* is more difficult but suggests those visitors who do not associate with either group and yet are emotionally involved. Therefore, visitors to a site might fall into any category or even more than one perhaps. This is indeed a relevant discussion, which requires more research in understanding the different types of visitors (*bystanders, victims, and perpetrators; non-tourists; tourists*), how they experience a site, and ultimately their actions/behaviour.

Specifically, it is claimed that even Auschwitz, the "spot that symbolizes the pinnacle of European dark tourism" (Tarlow, 2005, p. 45) and the "epitome of a dark tourism destination" (Stone & Sharpley, 2009, p. 587), being a must-visit tourist attraction provides an arena for various tourists and (*non tourists/non-visitors*) experiences, some have no dark elements to it. I would argue that nevertheless it does exist on two spectra viz. from light to dark and from weak to strong, which can, for example, include shame, curiosity, anger, or empathic grief. Accordingly, it can also be said that visitors to dark sites may experience dark or light. In Auschwitz, for example, you may have a visitor with a camp number tattoo (presumably experiencing thoughts of empathy with the victims), some army veterans (looking cheerful and experiencing pride, satisfaction, etc.), some bored Polish schoolchildren playing game (experiencing??), Israeli tourists waving flag of Israel (asserting a Jewish claim?), and two cheerful tourists taking their photo against the famous sign (experiencing a 'day out'?). The experience was different for each as was the heritage product they were consuming – for some various shades of dark, for others quite light. Thus, a camp or prison may evoke different experiences; for some it is dark, for others not. Note that while death may play a role in today's people's dark tourist experiences, its role will be limited, as death and dead bodies are re-entering the social consciousness and public realm, especially through the media and art (Giddens, 1991). Today's globolized TV shows as well as the news channels present dead bodies for view. Moreover, as almost everything can be observed in the safety and privacy of one's home via the internet, avoiding social danger, dark tourism will be based more on touching, smelling, and tasting. It is claimed that Larsen (2008) view that "tourist experience will move from the 'tourist gaze' and other representational approaches that privilege the eye" (Harwood & El-Manstrly, 2012), and will be "based more on being, doing, touching and seeing" (Holden, 2009) is specifically pertinent to dark tourism experience.

Emotions

Very little has been produced on the felt experience of dark tourism sites and their behavioural intentions, and importantly actions. The emotions that people have while they are visiting a site that is related to death will also give insight into their motivations for visiting those places. Nawijn and Fricke (2015) states the more emphasized the emotions are, the greater is the possibility for visitors to repeat the visit, and to recommend it to their friends and relatives. Positive and negative emotions indeed have received some consideration in dark tourism studies (Biran & Buda, 2018; Buda, 2015a; Nawijn & Biran, 2018; Nawijn, Isaac, van Liempt & Gridnesvskiy, 2016; Robinson & Picard, 2012; Tucker, 2009; Tucker & Shelton, 2018; Waterton & Watson, 2014).

To date only one researcher has focused on the affective dimensions of dark tourism (Buda, 2015a; b; Buda et al., 2014; Light, 2017, p. 288). Several authors (Lagos, Harris & Sigala, 2015; Hede & Hall, 2012) state there are still various literature gaps and academics call for more research aiming to better understanding the visitors' emotional experience (Prayag et al., 2017; Straker & Wringley, 2016). Hope, love, pride, fascination, interest are emotional experience that can also be segment of tourists' affective experiences (IIiev, 2020). Nawijn and Fricke (2015, p. 222) state that "existing empirical studies on emotional responses at sites associated with death and suffering is rather limited, as it is typically descriptive in nature". In addition, diverse emotions in conceptualization of the visitor experience at dark heritage sites cannot be overlooked (Oren, Shani & Poria, 2021; Zheng et al., 2019) along with the essential need to report positive contribution of negative emotions, whereas negative emotions contributed importantly to visitor's satisfaction (Oren, Shani & Poria, 2021).

References

Biran, A. & Buda, D.M. (2018) Unravelling fear of death motives in dark tourism. In P. Stone, R. Hartmann, A. Seaton, R. Sharpley & L. White (Eds.), *The Palgrave handbook of dark tourism studies* (pp. 515–532). London: Macmillan.

Biran, A. & Poria, Y. (2012) Re-conceptualising dark tourism. In R. Sharpley, & P. Stone (Eds.), *The contemporary tourism experience: Concepts and consequences* (pp. 62–79). London: Routledge.

Bowman, M.S. & Pezzulo, P.C. (2010) What's so 'dark' about 'dark tourism'?: Death, tours, and performance. *Tour Studies*, 9(3), 187–202.

Buda, D.M. (2015a) *Affective tourism: Dark routes in conflict*. New York: Routledge.

Buda, D.M. (2015b) The death drive in tourism studies. *Annals of Tourism Research*, 50, 39–51.

Buda, D.M., d'Hauteserre, A.-M. & Johnston, L. (2014). Feeling and tourism studies. *Annals of Tourism Research*, 46, 102–114.

Giddens, A. (1991). *Modernity and Self Identity*. Cambridge: Polity.

Harwood, S. & El-Manstrly, D. (2012). The performativity turn in tourism (University of Edinburgh Business School Working Paper Series, vol. 12/05). University of Edinburgh Business School.

Hede, A.-M. & Hall, J. (2012). Evoked emotions: Textual analysis within the context of pilgrimage tourism to Gallipoli. Advances in culture, *Tourism and Hospitality Research*, 6, 45–60.

Holden, A. (2009). The environment-tourism nexus: The influence of market ethics. *Annals of Tourism Research*, 36(3), 373–389.

Iliev, D. (2020). Consumption, motivation and experience in dark tourism: A conceptual and critical analysis. *Tourism Geographies*, 1–22. 23(5–6), 963–984.

Lagos, E., Harris, A. & Sigala, M. (2015). Emotional language for image formation and market segmentation in dark tourism destinations: Findings from tour operators' websites promoting Gallipoli. *Tourismos*, 10(2), 153–170.

Larsen, J. (2008). De-exoticizing tourist travel: Everyday life and sociality on the move. *Leisure Studies*, 27(1), 21–34.

Light, D. (2017). Progress in dark tourism and thanatourism research: An uneasy relationship with heritage tourism. *Tourism Management*, 61, 275–301. http://dx.doi.org/10.1016/j.tourman.2017.01.011

Nawijn, J. & Biran, A. (2018). Negative emotions in tourism: A meaningful analysis. *Current Issues in Tourism*, 22(19), 2386–2398.

Nawijn, J. & Fricke, M.-C. (2015). Visitor emotions and behavioural intentions: The case of Concentration Camp Memorial Neuengamme. *International Journal of Tourism Research*, 17(3), 221–228. http://dx.doi.org/10.1002/jtr.1977

Nawijn, J., Isaac, R.K., van Liempt, A. & Gridnevskiy, K. (2016). Emotion clusters for concentration camp memorials. *Annals of Tourism Research*, 61, 244–247.

Oren, G., Shani, A., & Poria, Y. (2021) Dialectical emotions in a dark heritage site: A study at the Auschwitz Death Camp. *Tourism Management*, 82, 104194 https://doi.org/10.1016/j.tourman.2020.104194

Prayag, G., Hosany, S., Muskat, B. & Del Chiappa, G. (2017). Understanding the relationships between tourists' emotional experiences, perceived overall image, satisfaction, and intention to recommend. *Journal of Travel Research*, 56(1), 41–54.

Robinson, M. & Picard, D. (2012). *Emoion in motion: Tourism, affect and transformation*. Farnham: Ashgate.

Stone, P. & Sharpley, R. (2009). Consuming dark tourism: A thanatological perspective. *Annals of Tourism Research*, 35(2), 574–595.

Straker, K. & Wrigley, C. (2016). Translating emotional insights into digital channel designs. *Journal of Hospitality and Tourism Technology*, 7(2), 135–157.

Tarlow, P. E. (2005). Dark tourism: The appealing 'dark' side of tourism and more. In M. Novelli (Ed.), *Niche tourism: Contemporary issues, trends and cases* (pp. 47–58). Amsterdam: Elsevier.

Tucker, H. (2009) Recognising emotions and its postcolonial potentialities: Discomfort, and shame in a tourism encounter in Turkey. *Tourism Geographies*, 11(4), 444–461.

Tucker, H. & Shelton, E. (2018) Tourism, mood and affect: Narratives of loss and hope. *Annals of Tourism Research*, 70, 66–75.

Waterton, E. & Watson, S. (2014). *The semiotic of heritage tourism*. Bristol: Channel View.

Zheng, C., Zhang, J., Qiu, M., Guo, Y. & Zhang, H. (2019). From mixed emotional experience to spiritual meaning: Learning in dark tourism places. *Tourism Geographies*, 22(1), 105–126.

Index